STRUCTURES AND SOLID BODY MECHANICS DIVISION

General Editor B. G. NEAL

THERMAL STRESS ANALYSES

THERMAL STRESS ANALYSES

D. J. JOHNS

Reader in Aeronautical Engineering, Loughborough College of Technology, Loughborough, Leicestershire
(Formerly, Lecturer in Aircraft Design, College of Aeronautics, Cranfield, Bletchley, Bucks.)

PERGAMON PRESS

OXFORD · LONDON · EDINBURGH · NEW YORK
PARIS · FRANKFURT

Pergamon Press Ltd., Headington Hill Hall, Oxford
4 & 5 Fitzroy Square, London W.1

Pergamon Press (Scotland) Ltd., 2 & 3 Teviot Place, Edinburgh 1

Pergamon Press Inc., 122 East 55th Street, New York 22, N.Y.

Pergamon Press GmbH, Kaiserstrasse 75, Frankfurt-am-Main

Federal Publications Ltd., Times House, River Valley Rd., Singapore

Samcax Book Services Ltd., Queensway, P.O. Box 2720, Nairobi, Kenya

First edition 1965

Library of Congress Catalog Card No. 65-18374

Set in 10 on 12pt. Baskerville
and printed Great Britain
by Bell and Bain Ltd., Glasgow

CONTENTS

PREFACE

THERMAL stress problems occur in many branches of engineering and have already received considerable attention both in analysis and design. Such stresses result when differential thermal expansions are caused in a solid body, and if these stresses are high and associated with high temperatures in the body the yield stress of the material at these temperatures may be approached or even exceeded. It may be seen therefore that thermal stress problems can arise in any branch of engineering where temperature gradients are possible, and the only essential difference between the various problems is in the nature of the process causing the temperature gradient.

Thus, propulsive systems are inherently prone to high temperature effects resulting from the combustion process, whilst in the nuclear and aeronautical fields extremely high temperatures, and large gradients, can occur due to the fission process and the phenomena of aerodynamic heating respectively. In a recent study of ship failures thermal stresses of moderate to severe intensity were found to be present in more than one-third of those studied; the causes could be solar radiation, the heating of fuel oil or the cooling of refrigerated spaces.

The problems which arise fall naturally into three categories. The first comprises the problems accompanying a comparatively small, uniform rise in temperature; the thermal stresses caused are negligible and in a stress analysis due allowance must be made for variations in material properties with temperature. The second category is the main concern of the present book and deals with large variations in temperature and thermal expansion which produce thermal stresses (elastic or plastic) in materials whose properties are time-independent. The third category involves materials and conditions which may introduce significant time-dependent effects such as creep. This last category involves many complex problems and consideration must be given in any structural analysis to such effects as the redistribution of stress due to creep, creep buckling, etc.

The decision to omit consideration of problems in the third category from this book, and to concentrate on the second category is, of course, regretted, but the author feels that in a book of this size it is preferable to give greater attention to a smaller class of problems.

The restriction on length also prevents the presentation of more solutions to actual problems, but the author hopes that the many references given will enable the reader to pursue quickly, his subject of interest. In this connexion the two, most extensive, bibliographies listed at the end of the References contain between them over 800 references for the period up to 1963.

ACKNOWLEDGEMENTS

ACKNOWLEDGEMENT is made to the following publishers who gave permission for figures to be reproduced in this book ;

Advisory Group for Aeronautical Research and Development.

Aeronautical Research Institute, Univ. of Tokyo.

Aeronautical Society of India.

American Institute of Aeronautics and Astronautics.

Her Majesty's Stationery Office.

Royal Aeronautical Society.

Society of Automotive Engineers.

Weizmann Science Press (Israel).

My thanks are also due to Professor B. G. Neal, Imperial College of Science and Technology, who invited me to write the book ; to Professor G. M. Lilley, College of Aeronautics, Cranfield, who made comprehensive comments and suggestions on the contents of the Appendix, and to Professor W. S. Hemp, College of Aeronautics, Cranfield, for his encouragement during the preparation of the manuscript and his general comments on its final form. For her enthusiasm and skill my most sincere thanks go to Mrs. J. Carberry who typed and prepared the manuscript.

D. J. JOHNS

College of Aeronautics, Cranfield, and
Loughborough College of Technology

December 1963/*January* 1964

PRINCIPAL NOTATION

$a, b, c,$	linear dimensions
A	area
B	Biot Number ($= hd/k$) ; differential operator in Chapter 6
C	heat capacity per unit volume
d	thickness
D	flexural rigidity ($= Ed^3/12(1 - \nu^2)$)
e ; e_{ij}	strain
$\bar{e}$	dilatation ($= e_{xx} + e_{yy} + e_{zz}$)
E ; E_s	Young's Modulus ; secant modulus
F	free-energy function ; resultant vertical load on shell
G	shear modulus
G_i	Gibbs function
h	heat transfer coefficient
H	internal heat generation rate
i	enthalpy
I	second moment of area
J	mechanical equivalent of heat
k	thermal conductivity
K ; $\bar{K}$	bulk modulus ($= E/3(1 - 2\nu)$), spring stiffness ; flexibility
l, m, n	direction cosines
L	length ; potential energy
L'	complementary potential energy
M_x, M_y, M_{xy}	bending moment resultants
M_a	Mach Number of free stream flow
M_T	$E\alpha \int Tz dz$
N_x, N_y, N_{xy}	force resultants
N_T	$E\alpha \int T dz$
p	pressure
P	load
q	heat flux per unit area per unit time
Q	heat supplied per unit volume
Q_x, Q_y	shear force
r, θ, z	cylindrical coordinate system
r, θ, ϕ	spherical coordinate system
r_b	b_w/b_s ($=$ ratio of web depth to skin width)

r_t	t_w/t_s (= ratio of web thickness to skin thickness)
R	radius of curvature
R, Θ, Z	body forces per unit volume in cylindrical coordinates
R, Θ, Φ	body forces per unit volume in spherical coordinates
R_c	concentrated reaction
S	Reynolds analogy factor
t	time; thickness
t_w, t_s	web thickness; skin thickness
T	temperature rise above initial stress-free state
T_o	temperature at initial stress-free state (absolute)
T'	absolute temperature ($= T + T_0$)
u, v, w	components of displacement vector in coordinate directions.
U	intrinsic energy per unit volume
U_o	strain energy
$\bar{U}_o$	complementary strain energy
U_a	free-stream velocity
U, V	variables introduced in shell theory
W	work done per unit volume; parameter in heat transfer ($= \kappa t/d^2$)
W_u	loss in potential energy of surface forces
W_t	loss in potential energy of body forces
x, y, z	cartesian coordinate system
X, Y, Z	body forces per unit volume in rectangular cartesian coordinates
X_s, Y_s, Z_s	surface forces per unit area in rectangular cartesian coordinates
α	coefficient of linear thermal expansion
β	$E\alpha/(1 - 2\nu)$; shell parameter $[3(1 - \nu^2)/d^2R^2]^{1/4}$
δ	displacement
ϵ	emissivity
η	entropy per unit volume
θ	rotation; shell angular coordinate
κ	thermal diffusivity; curvature change
λ, μ	Lamé's elastic constants (Chapter 1)
ν	Poisson's ratio
ρ	mass density; least radius of gyration
σ, σ_{ij}	stress

Σ	$\sigma_{xx} + \sigma_{yy} + \sigma_{zz}$
τ	twist
ϕ	Airy stress function ; beam rotation
ϕ_N, ϕ_T	force functions
ψ	potential function ; shell angular coordinate

CHAPTER 1

Fundamentals of Thermal Stress Analysis

1.1. Preliminary Remarks on Thermal Stress

Most substances expand when their temperature is raised and contract when cooled, and for a wide range of temperatures this expansion or contraction is proportional to the temperature change. This proportionality is expressed by the coefficient of linear thermal expansion (α) which is defined as the change in length which a bar of unit length undergoes when its temperature is changed by 1°.

If free expansion or contraction of all the fibres of a body is permitted, no stress is caused by the change in temperature. However, when the temperature rise in a homogeneous body is not uniform, different elements of the body tend to expand by different amounts and the requirement that the body remain continuous conflicts with the requirement that each element expand by an amount proportional to the local temperature rise. Thus the various elements exert upon each other a restraining action resulting in continuous unique displacements at every point. The system of strains produced by this restraining action cancels out all, or part of, the free thermal expansions at every point so as to ensure continuity of displacement. This system of strains must be accompanied by a corresponding system of self-equilibrating stresses. These stresses are known as **thermal stresses.**

A similar system of stresses may be induced in a structure made of dissimilar materials even when the temperature change throughout the structure is uniform. Also, if the temperature change in a homogeneous body is uniform and external restraints limit the amount of expansion or contraction, the stresses produced in the body are termed **thermal stresses.**

These various ideas can be illustrated by the following simple example. Two parallel rods of different materials and lengths are fixed at one end and are restrained to move together at their other end, see Fig. 1.1. Movement of the combined structure is only permitted in the direction parallel to the rod axes, and it is reacted by an elastic spring of stiffness K. If T_1 and T_2 denote the rise in

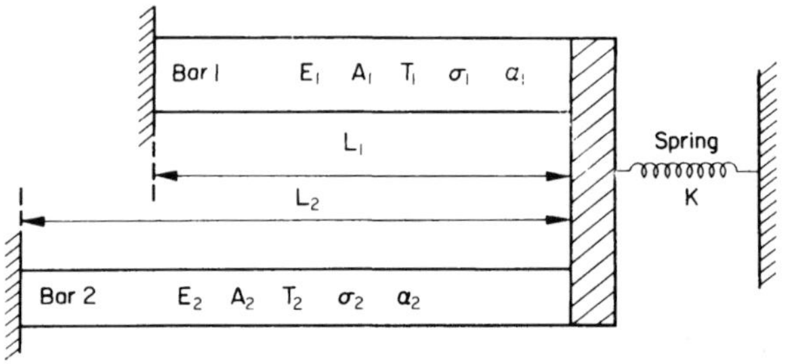

FIG. 1.1. Simple two-bar structure.

temperature from the initial stress-free state, experienced by each rod, the conditions of equilibrium and compatibility of strain are given, respectively, by

$$\sigma_1 A_1 + \sigma_2 A_2 = P \tag{1.1}$$

and

$$\alpha_1 T_1 L_1 + \frac{\sigma_1 L_1}{E_1} = \alpha_2 T_2 L_2 + \frac{\sigma_2 L_2}{E_2} = -\delta \tag{1.2}$$

Here, σ_1 and σ_2 are the corresponding tensile thermal stresses in the two rods, of cross-sectional areas A_1 and A_2 respectively, and P is the compressive load in the spring which has a compressive displacement δ (i.e. $P = K\delta$). In eqn. (1.2) the terms αTL refer to the free thermal expansions whilst the terms σ/E refer to the strains necessary to ensure continuity of displacement (E is the modulus of elasticity).

The solution of the above equations yields the following result for σ_1

$$\sigma_1\left[1+\frac{KL_1}{A_1E_1}+\frac{A_2E_2L_1}{A_1E_1L_2}\right]$$
$$=\frac{-A_2E_2}{A_1L_2}\left[a_1T_1L_1-a_2T_2L_2\right]-\frac{Ka_1T_1L_1}{A_1} \qquad (1.3)$$

and the corresponding result for σ_2 can be obtained by substitution of (1.3) into either (1.1) or (1.2).

If $K = 0$ the spring is infinitely flexible, the combined structure is unrestrained and the expression for σ_1 becomes

$$\sigma_1 = -\frac{E_1}{L_1}\left[a_1T_1L_1 - a_2T_2L_2\right]\Big/\left[1+\frac{A_1E_1L_2}{A_2E_2L_1}\right] \qquad (1.4)$$

If $K = \infty$ the structure is completely restrained against axial displacement, and

$$\sigma_1 = -E_1a_1T_1 \qquad (1.5)$$

From (1.4) it can be deduced that for a composite structure made of dissimilar materials, and with a non-uniform temperature rise the self-equilibrating thermal stresses depend not only upon the temperature differences in the structure, but also upon the geometry (A, L) of the various structural elements. It is also seen that, if the unrestrained structure is homogeneous, thermal stresses result from non-uniform temperatures, and even for uniform temperatures thermal stresses can occur in a non-homogeneous structure.

From (1.5) it can be deduced that in a fully restrained element the thermal stress magnitude depends only upon the physical properties and temperature rise of the element concerned. For the case of an element which is fully restrained in all three mutually perpendicular directions the thermal stress would be (see eqn. (1.50))

$$\sigma = -EaT/(1-2\nu) \qquad (1.6)$$

which for $\nu = 0{\cdot}3$ would be

$$\sigma = -2{\cdot}5EaT \qquad (1.7)$$

One can in general represent the maximum thermal stress as $BEaT$ and for temperature distributions which change slowly with time, so that there are no significant dynamic effects, the maximum value of B would be 2·5, as shown above, for a fully restrained homogeneous body. When dynamic effects are present B may be higher, and the effective value of B may also be increased by the presence of holes, notches or other stress raisers. The product $E\alpha$ for a material is known as the thermal stress modulus and, as can be seen in the examples above, its value is a measure of the maximum thermal stress possible in a homogeneous body. Typical values are 280 lb/in^2 °C for steel and 150 lb/in^2 °C for titanium so that significantly high thermal stresses are obtained for a temperature rise of, say, 100°C in a fully restrained structure.

It should be remembered that the thermal stresses are influenced by the temperature level because the material properties are generally temperature-dependent and the body may then behave as a non-homogeneous, elastic, or elastic–plastic medium. However, for many cases of practical interest it is reasonable to assume that the material is elastic and perform thermal stress calculations using uniform elastic properties corresponding to the average temperature.

In the simple analysis above, the thermal stresses were calculated assuming that the temperature distribution was known and only simple mechanical concepts were used in the analysis. However, in any exact analysis due allowance should be taken of the appropriate concepts of *mechanics* and *thermodynamics*.

A thorough treatment of the thermodynamic aspects of the problem has been presented by Boley and Weiner in their excellent book(15) and the reader seeking a complete fundamental development of the subject should refer to that text. For the present volume an exposition of the main results is all that is needed and the theorems, principles and formulae subsequently presented in this chapter are based mainly on a paper by Hemp(63) which relies on previous work by Love(103) and Roberts(145). To anticipate later results it will be shown (Section 1.6) that the desired energy equation for an isotropic elastic solid is

$$k\nabla^2 T = (C)_{e=0}\frac{\partial T}{\partial t} + \beta T_0 \frac{\partial \bar{e}}{\partial t}, \tag{1.8}$$

where T is the temperature rise from the initial uniform temperature T_0, of the stress-free state; k is the thermal conductivity; t is the time; $(C)_{e=0}$ is the heat capacity per unit volume at zero strain; $\bar{e}$ is the dilatation and $\beta = E\alpha/(1-2\nu)$.

This equation is based on the Fourier law of heat conduction and the linear thermoelastic stress–strain relations, and it shows that the temperature distribution in a body depends upon the dilatations $\bar{e}$ throughout the body. Thus, the temperature and strain (and, hence, stress) distributions are coupled and an exact analysis would require the *simultaneous* determination of the stress and temperature distributions.[15] Fortunately this thermoelastic coupling effect is small and it will not be considered in any of the thermal stress analyses in this volume.

Equation (1.8) was originally presented by Duhamel in 1835 in a paper read before the Academy of Sciences. In this[30] and subsequent papers[29, 31] he modified the equations of isothermal elasticity presented by Poisson[138] in 1829 and so established the analytical foundations of thermoelastic theory; he examined the implications of various simplifications and presented solutions to several specific problems. Almost simultaneously the thermoelastic stress equations were derived also by Neumann[125].

These equations will be presented later in their general three-dimensional form for a continuous, solid, isotropic body, assumed to be subject to body forces (e.g. X, Y, Z) per unit volume and surface forces (X_s, Y_s, Z_s) per unit surface area. The heat transfer boundary conditions are assumed known. Variations with time of the thermal and mechanical loading are admissible, and if these are so rapid that vibrations are excited in the body, then the body forces are considered to include the inertia forces. Thermoelastic coupling is specifically excluded from further consideration however, so that the temperature distributions can be determined independently of the deformations, and both can be calculated separately as functions of time. The determination of the temperature distributions in a structure and general considerations of heat transfer in structures are both given in the Appendix.

1.2. Definition of Strain Components

As in isothermal elasticity the strain–displacement relations are derived directly from purely geometrical considerations. The displacements are assumed to be small and are taken from an initial unstressed condition at uniform temperature, T_0. The displacements are assumed not to affect the body geometry or the density.

The equations for a rectangular axis system (x, y, z) are:

$$\left.\begin{aligned} e_{xx} &= \frac{\partial u}{\partial x}\,; \quad e_{yy} = \frac{\partial v}{\partial y}\,; \quad e_{zz} = \frac{\partial w}{\partial z} \\ e_{xy} &= \frac{\partial u}{\partial y} + \frac{\partial v}{\partial x}\,; \quad e_{yz} = \frac{\partial v}{\partial z} + \frac{\partial w}{\partial y}\,; \quad e_{zx} = \frac{\partial w}{\partial x} + \frac{\partial u}{\partial z} \end{aligned}\right\} \tag{1.9}$$

where u, v and w are the components of the displacement vector in the x, y, and z directions respectively.

The corresponding results for a cylindrical coordinate system (r, θ, z) are:

$$\left.\begin{aligned} e_{rr} &= \frac{\partial u}{\partial r}\,; \quad e_{\theta\theta} = \frac{u}{r} + \frac{1}{r}\frac{\partial v}{\partial \theta}\,; \quad e_{zz} = \frac{\partial w}{\partial z} \\ e_{r\theta} &= \frac{1}{r}\frac{\partial u}{\partial \theta} + \frac{\partial v}{\partial r} - \frac{v}{r}\,; \quad e_{\theta z} = \frac{\partial v}{\partial z} + \frac{1}{r}\frac{\partial w}{\partial \theta}\,; \quad e_{zr} = \frac{\partial w}{\partial r} + \frac{\partial u}{\partial z} \end{aligned}\right\} \tag{1.10}$$

and in spherical coordinates (r, θ, ϕ):

$$\left.\begin{aligned} e_{rr} &= \frac{\partial u}{\partial r}\,; \quad e_{\theta\theta} = \frac{u}{r} + \frac{1}{r}\frac{\partial v}{\partial \theta}\,; \quad e_{\phi\phi} = \frac{u}{r} + \frac{v}{r}\cot\theta + \frac{1}{r\sin\theta}\frac{\partial w}{\partial \phi} \\ e_{r\theta} &= \frac{1}{r}\frac{\partial u}{\partial \theta} + \frac{\partial v}{\partial r} - \frac{v}{r}\,; \quad e_{\theta\phi} = \frac{1}{r}\frac{\partial w}{\partial \theta} - \frac{\cot\theta}{r}w + \frac{1}{r\sin\theta}\frac{\partial v}{\partial \phi} \\ e_{\phi r} &= \frac{\partial w}{\partial r} - \frac{w}{r} + \frac{1}{r\sin\theta}\frac{\partial u}{\partial \phi} \end{aligned}\right\} \tag{1.11}$$

These various systems are shown in Figs. 1.2 and 1.3 respectively.

From a physical point of view the displacements in a simply connected body must be single-valued and continuous. Certain restrictions on the strains e_{ij} arise to meet this requirement and these constitute the so-called strain compatibility equations. For whichever coordinate system is being used the six strains e_{ij} are written in terms of the displacements u, v, and w and by repeated differentiation and elimination of displacements the appropriate equations of compatibility are derived which are valid for both thermal and mechanical loading.

Thus, in rectangular coordinates,

$$\left.\begin{aligned}
&\frac{\partial^2 e_{xx}}{\partial y^2}+\frac{\partial^2 e_{yy}}{\partial x^2}=\frac{\partial^2 e_{xy}}{\partial x\partial y}\,;\quad \frac{\partial^2 e_{yy}}{\partial z^2}+\frac{\partial^2 e_{zz}}{\partial y^2}=\frac{\partial^2 e_{yz}}{\partial y\partial z}\\
&\frac{\partial^2 e_{zz}}{\partial x^2}+\frac{\partial^2 e_{xx}}{\partial z^2}=\frac{\partial^2 e_{xz}}{\partial x\partial z}\,;\quad \frac{2\partial^2 e_{xx}}{\partial y\partial z}=\frac{\partial}{\partial x}\left(\frac{\partial e_{zx}}{\partial y}+\frac{\partial e_{xy}}{\partial z}-\frac{\partial e_{yz}}{\partial x}\right)\\
&\frac{2\partial^2 e_{yy}}{\partial z\partial x}=\frac{\partial}{\partial y}\left(\frac{\partial e_{xy}}{\partial z}+\frac{\partial e_{yz}}{\partial x}-\frac{\partial e_{zx}}{\partial y}\right);\quad \frac{2\partial^2 e_{zz}}{\partial x\partial y}\\
&\qquad\qquad=\frac{\partial}{\partial z}\left(\frac{\partial e_{yz}}{\partial x}+\frac{\partial e_{zx}}{\partial y}-\frac{\partial e_{xy}}{\partial z}\right)
\end{aligned}\right\}\quad(1.12)$$

If the body is multiply connected (i.e. has internal cavities) then additional conditions are necessary to ensure single valued displacements. These conditions specify that the limiting values of the

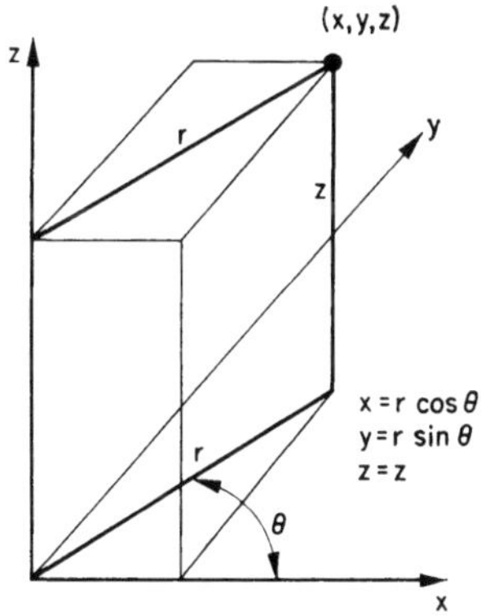

FIG. 1.2. Cylindrical coordinate system.

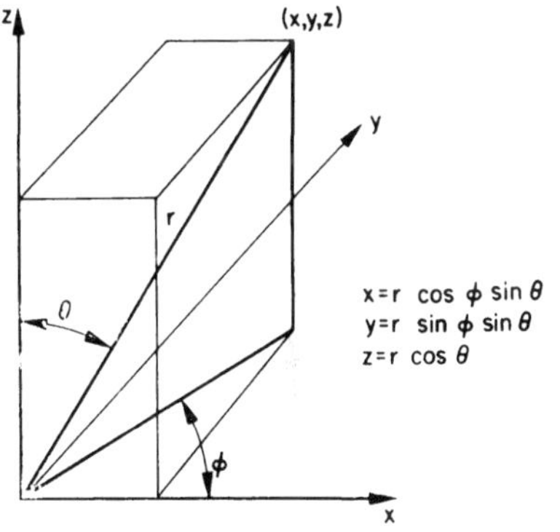

Fig. 1.3. Spherical coordinate system.

displacements at imaginary cuts which make the body simply-connected should be the same when a cut is approached from either side.[15]

1.3. Equations of State

From the remarks in Section 1.1 it is seen that the total strains at each point in a heated body may be considered to consist of two parts, viz. the free thermal expansion plus the strains dependent upon the stress state in the body. Since the first part is uniform in all directions at a given point in an isotropic body it can be deduced that no shear strains result—only direct strains. Therefore, the usual strain–stress relations of linear isothermal elasticity are extended for the thermoelastic case to become,

$$\left.\begin{aligned} e_{xx} &= \alpha T + [\sigma_{xx} - \nu(\sigma_{yy} + \sigma_{zz})]/E \\ e_{yy} &= \alpha T + [\sigma_{yy} - \nu(\sigma_{zz} + \sigma_{xx})]/E \\ e_{zz} &= \alpha T + [\sigma_{zz} - \nu(\sigma_{xx} + \sigma_{yy})]/E \\ e_{xy} &= \sigma_{xy}/G \; ; \; e_{yz} = \sigma_{yz}/G \; ; \; e_{zx} = \sigma_{zx}/G \end{aligned}\right\} \quad (1.13)$$

Solution of (1.13) yields

$$\left.\begin{aligned} \sigma_{xx} &= (\lambda + 2\mu)e_{xx} + \lambda(e_{yy} + e_{zz}) - \beta T \\ \sigma_{yy} &= (\lambda + 2\mu)e_{yy} + \lambda(e_{zz} + e_{xx}) - \beta T \\ \sigma_{zz} &= (\lambda + 2\mu)e_{zz} + \lambda(e_{xx} + e_{yy}) - \beta T \\ \sigma_{xy} &= \mu e_{xy} \; ; \; \sigma_{yz} = \mu e_{yz} \; ; \; \sigma_{zx} = \mu e_{zx} \end{aligned}\right\} \quad (1.14)$$

where λ and μ, the Lamé elastic constants, are defined by

$$\lambda = \nu E/(1 + \nu)\,(1 - 2\nu) \; ; \; \mu = E/2(1 + \nu) = G \; ; \text{ and } \beta = E\alpha/(1 - 2\nu) \quad (1.15)$$

Equations (1.13) or (1.14) are the equations of state for the elastic regime of an isotropic solid body. However, the linear relations of (1.13) can be modified for non-linear elasticity by employing an equivalent secant modulus E_s in place of E, where

$$E_s = \sigma_{ii}/(e_{ii} - \alpha T) \quad (1.16)$$

Unfortunately E_s will not be constant and direct solution of problems will be difficult. A progressive solution considering small increments of load–time history must assume the sense of the incremental stress since a material may have different moduli for different directions of loading. This aspect will be considered further in Chapter 8.

It should be noted that E, G, ν and α (and hence, λ, μ and β) are all, in general, functions of temperature.

If the dilatation $\bar{e}$ and the sum of the normal stresses Σ are defined by

$$\bar{e} = e_{xx} + e_{yy} + e_{zz} \tag{1.17}$$

and

$$\Sigma = \sigma_{xx} + \sigma_{yy} + \sigma_{zz} \tag{1.18}$$

the following relation is obtained from (1.13)

$$\bar{e} = 3\alpha T + \Sigma/3K \tag{1.19}$$

where K is the bulk modulus defined by

$$K = E/3(1 - 2\nu) \tag{1.20}$$

1.4. Equations of Equilibrium

The components of stress must satisfy the usual equations of equilibrium throughout the volume, and since these are derived from purely mechanical considerations they are the same as those for isothermal elasticity,(163) e.g. in rectangular cartesian coordinates (x, y, z)

$$\left.\begin{aligned} \frac{\partial \sigma_{xx}}{\partial x} + \frac{\partial \sigma_{xy}}{\partial y} + \frac{\partial \sigma_{xz}}{\partial z} + X = 0 \\ \frac{\partial \sigma_{xy}}{\partial x} + \frac{\partial \sigma_{yy}}{\partial y} + \frac{\partial \sigma_{yz}}{\partial z} + Y = 0 \\ \frac{\partial \sigma_{xz}}{\partial x} + \frac{\partial \sigma_{yz}}{\partial y} + \frac{\partial \sigma_{zz}}{\partial z} + Z = 0 \end{aligned}\right\} \tag{1.21}$$

with the complementary shear stress components equal,

i.e. $$\sigma_{xy} = \sigma_{yx},\ \sigma_{yz} = \sigma_{zx},\ \sigma_{zx} = \sigma_{xz} \tag{1.22}$$

As mentioned earlier, if inertia effects are to be considered then, for small displacements, with ρ being the mass density

$$X = -\rho \frac{\partial^2 u}{\partial t^2},\quad Y = -\rho \frac{\partial^2 v}{\partial t^2},\quad Z = -\rho \frac{\partial^2 w}{\partial t^2} \tag{1.23}$$

Gravity forces may be included similarly.

The corresponding equations in cylindrical coordinates (r, θ, z) are

$$\left.\begin{aligned}
&\frac{\partial \sigma_{rr}}{\partial r} + \frac{1}{r}\frac{\partial \sigma_{r\theta}}{\partial \theta} + \frac{\partial \sigma_{rz}}{\partial z} + \frac{\sigma_{rr} - \sigma_{\theta\theta}}{r} + R = 0 \\
&\frac{\partial \sigma_{r\theta}}{\partial r} + \frac{1}{r}\frac{\partial \sigma_{\theta\theta}}{\partial \theta} + \frac{\partial \sigma_{\theta z}}{\partial z} + \frac{2\sigma_{r\theta}}{r} + \Theta = 0 \\
&\frac{\partial \sigma_{rz}}{\partial r} + \frac{1}{r}\frac{\partial \sigma_{z\theta}}{\partial \theta} + \frac{\partial \sigma_{zz}}{\partial z} + \frac{\sigma_{rz}}{r} + Z = 0
\end{aligned}\right\} \tag{1.24}$$

where R, Θ and Z are the corresponding body force components; and in spherical coordinates (r, θ, ϕ) the equations have the form

$$\left.\begin{aligned}
&\frac{\partial \sigma_{rr}}{\partial r} + \frac{1}{r}\frac{\partial \sigma_{r\theta}}{\partial \theta} + \frac{1}{r\sin\theta}\frac{\partial \sigma_{r\phi}}{\partial \phi} \\
&\qquad + \frac{1}{r}\left[2\sigma_{rr} - \sigma_{\theta\theta} - \sigma_{\phi\phi} + \sigma_{r\theta}\cot\theta\right] + R = 0 \\
&\frac{\partial \sigma_{r\theta}}{\partial r} + \frac{1}{r}\frac{\partial \sigma_{\theta\theta}}{\partial \theta} + \frac{1}{r\sin\theta}\frac{\partial \sigma_{\theta\phi}}{\partial \phi} \\
&\qquad + \frac{1}{r}\left[\cot\theta\,(\sigma_{\theta\theta} - \sigma_{\phi\phi}) + 3\sigma_{r\theta}\right] + \Theta = 0 \\
&\frac{\partial \sigma_{r\phi}}{\partial r} + \frac{1}{r}\frac{\partial \sigma_{\theta\phi}}{\partial \theta} + \frac{1}{r\sin\theta}\frac{\partial \sigma_{\phi\phi}}{\partial \phi} \\
&\qquad + \frac{1}{r}\left[3\sigma_{r\phi} + 2\cot\theta\,\sigma_{\theta\phi}\right] + \Phi = 0
\end{aligned}\right\} \tag{1.25}$$

where R, Θ and Φ are the corresponding body force components.

1.5. Boundary Conditions

1.5.1. Traction Boundary Conditions

Not only must the components of stress satisfy the equations of equilibrium throughout the volume of the body but also at all points on the surface, i.e.

$$\left.\begin{aligned}
l\sigma_{xx} + m\sigma_{xy} + n\sigma_{xz} &= X_s \\
l\sigma_{xy} + m\sigma_{yy} + n\sigma_{yz} &= Y_s \\
l\sigma_{xz} + m\sigma_{yz} + n\sigma_{zz} &= Z_s
\end{aligned}\right\} \tag{1.26}$$

where l, m and n are the direction cosines of the outward drawn normal.

1.5.2. Displacement Boundary Conditions

The components of the displacement vector must satisfy the conditions below at each point, s, on the surface,

$$u = f_1(s), \quad v = f_2(s), \quad w = f_3(s) \tag{1.27}$$

where f_1, f_2 and f_3 are prescribed functions.

1.5.3. Mixed Boundary Conditions

In practice of course it is possible that neither of the above sets of conditions would apply independently over the entire surface of the body, and mixed boundary conditions are then involved. To generalize therefore, at each point on the surface any three of the six conditions above may be specified provided that each of the three is related to a different coordinate direction, e.g. u, v and Z_s may be specified but not u, v and Y_s.

It is also possible to prescribe the condition whereby a relation exists between a corresponding pair of displacement and surface force components. This was assumed in the example in Section 1.1 where the axial displacement δ and the spring load P were related through the spring constant, K.

1.6. Thermodynamic Considerations and Thermoelastic Coupling[63]

At any point in the body the macroscopic state is determined by the strain components and the temperature. The stress components, entropy η per unit volume and the intrinsic energy U per unit volume are all functions of these variables. The First Law of Thermodynamics may be formulated for a small change of state as follows,

$$\delta U = \delta Q + \delta W \tag{1.28}$$

where δQ is the heat supplied per unit volume to the neighbourhood of a point in the body and δW is the work done per unit volume by the surrounding material.

From the Second Law of Thermodynamics, the relation

$$\delta Q = (T_0 + T)\,\mathrm{d}\eta = T'\,\mathrm{d}\eta \tag{1.29}$$

is obtained, which is valid if the processes are quasi-static and the state of the body is sensibly uniform in the neighbourhood of a point. Using previously defined notation it can be shown that

$$\delta W = \sigma_{xx}\,\delta e_{xx} + \ldots + \sigma_{xy}\,\delta e_{xy} + \ldots \tag{1.30}$$

and hence

$$\delta U = T'\,\delta\eta + \sigma_{xx}\,\delta e_{xx} + \ldots + \sigma_{xy}\,\delta e_{xy} + \ldots \tag{1.31}$$

In these equations T' is defined as the absolute temperature or

$$T' = T + T_0 \tag{1.32}$$

where T is the temperature rise from the initial uniform temperature T_0, of the stress-free state.

The " free energy " F per unit volume is defined by[145]

$$F = U - T'\eta \tag{1.33}$$

and is regarded as a function of the strains and temperature.

The " thermodynamic potential " or " Gibbs function " G_i per unit volume is defined as[145]

$$G_i = U - T'\eta - \sigma_{xx}e_{xx} - \ldots - \sigma_{xy}e_{xy} - \ldots \tag{1.34}$$

and is regarded as a function of the stresses and temperature.

The expressions for F and G_i as derived by Hemp[63] are

$$F = \frac{1}{2}\left\{(\lambda + 2\mu)\bar{e}^2 + \mu(e_{xy}^2 + e_{yz}^2 + e_{zx}^2 - 4e_{xx}e_{yy} - 4e_{yy}e_{zz} - 4e_{zz}e_{xx})\right\} - \beta T\bar{e} + C_1(T') \tag{1.35}$$

$$-G_i = \frac{1}{2E}\left[\Sigma^2 + 2(1+\nu)\left\{\sigma_{xy}^2 + \sigma_{yz}^2 + \sigma_{zx}^2 - \sigma_{xx}\sigma_{yy} - \sigma_{yy}\sigma_{zz} - \sigma_{zz}\sigma_{xx}\right\}\right] + \alpha T\Sigma + C_2(T') \tag{1.36}$$

where both C_1 and C_2 are functions of T'.

From (1.31) and (1.33) it follows that

$$\frac{\partial F}{\partial T'} = -\eta\ ; \quad \frac{\partial F}{\partial e_{ij}} = \sigma_{ij} \tag{1.37}$$

and from (1.31) and (1.34), similarly,

$$\frac{\partial G_i}{\partial T'} = -\eta\ ; \quad \frac{\partial G_i}{\partial \sigma_{ij}} = -e_{ij} \tag{1.38}$$

where the corresponding expressions for entropy η are functions of temperature and the strains e_{ij} and stresses σ_{ij} respectively. The heat capacity per unit volume at zero strain $(C)_{e=0}$ can be defined from (1.29) as

$$(C)_{e=0} = T'\left(\frac{\partial \eta}{\partial T'}\right)_{e=0} \tag{1.39}$$

Therefore from (1.35) and (1.37)

$$(C)_{e=0} = -T'\left(\frac{\partial^2 F}{\partial T'^2}\right)_{e=0} \tag{1.40}$$

and hence

$$C_1 = -\int_0^T\!\!\int_0^T \frac{(C)_{e=0}}{T'}\,\mathrm{d}T\mathrm{d}T \tag{1.41}$$

If similarly the heat capacity per unit volume at zero stress is defined as

$$(C)_{\sigma=0} = T\left(\frac{\partial \eta}{\partial T'}\right)_{\sigma=0}$$

then a corresponding expression for C_2 to (1.41) is obtained with $(C)_{\sigma=0}$ instead of $(C)_{e=0}$. The relevance of the terms F and G_i will be shown in the next section, whilst the significance of the entropy η to the question of thermoelastic coupling will now be discussed.

It is shown in the Appendix that the thermal conduction problem requires the solution of eqn. (A.6), viz.

$$T'\frac{\partial \eta}{\partial T} = \frac{\partial}{\partial x}\left(k\frac{\partial T}{\partial x}\right) + \frac{\partial}{\partial y}\left(k\frac{\partial T}{\partial y}\right) + \frac{\partial}{\partial z}\left(k\frac{\partial T}{\partial z}\right) \tag{1.42}$$

where k is the thermal conductivity of the material comprising the solid body. In general the thermal conduction problem is linked through η to the state of strain or stress, e.g. eqns. (1.37) and (1.35).

Thus, writing $\frac{\partial T'}{\partial t}$ as $\dot{T}'$, etc,

$$T'\frac{\partial \eta}{\partial t} = T'\left[\frac{\partial \eta}{\partial e_{ij}}\dot{e}_{ij} + \frac{\partial \eta}{\partial T'}\dot{T}'\right]$$

$$= -\,T'\left[\frac{\partial^2 F}{\partial e_{ij}\,\partial T'}\dot{e}_{ij} + \frac{\partial^2 F}{\partial T'^2}\dot{T}'\right]$$

Using (1.40), then (1.37) and (1.14)

$$T'\frac{\partial \eta}{\partial t} = (C)_{e=0}\,\dot{T}' - T'\dot{e}_{ij}\frac{\partial^2 F}{\partial e_{ij}\,\partial T'}$$

$$= (C)_{e=0}\ddot{T}' - T'\dot{e}_{ij}\frac{\partial \sigma_{ij}}{\partial T'}$$

$$= (C)_{e=0}\dot{T}' + T'\dot{\bar{e}}\beta \tag{1.43}$$

If a linear result is desired in (1.43) it may be assumed that the dilatation, $\bar{e}$, and the temperature rise, T, are sufficiently small so that their product may be neglected. Thus the terms to be substituted into the left-hand side of (1.42) are

$$(C)_{e=0}\dot{T} + T_0\dot{\bar{e}}\beta,$$

and this equation (1.42) together with (1.13) forms the basis of thermoelastic coupling in that both the strains e_{ij} and the temperature T are dependent upon one another in an exact solution. Fortunately the term in $\dot{\bar{e}}$ above can usually be neglected, as can any distinction between $(C)_{e=0}$ and $(C)_{\sigma=0}$. Writing these terms simply as C, the heat capacity per unit volume, results in the well-known form of the uncoupled heat conduction equation, (A.7).

1.7. Minimal Principles in Thermoelasticity

Energy principles, which yield relations between the stresses, strains, forces and displacements when adequate conditions are specified, may be used to describe the behaviour of a structure.[159] Typical conditions would be those on the bounding surface, in addition to those specifying the temperature–load–time history, material properties, etc.

The Theorems of Stationary Potential Energy and Stationary Complementary Potential Energy as used in isothermal elasticity have similar applications in thermoelastic problems. The potential energy is defined as

$$L = \bar{U}_o - W_t$$

and the complementary potential energy as,

$$L' = \bar{U}'_o - W_u$$

where $\bar{U}_o$ is the strain energy of the body, and $\bar{U}'_o$ is the complementary strain energy. W_t is the loss in potential energy of the body forces over the total volume (assuming the body forces are prescribed over the total volume) and of the surface tractions over those portions of the surface where the tractions are prescribed. W_u is the loss in potential energy of the surface tractions over those portions of the surface where the displacements and not the tractions are prescribed. The complementary strain energy in this case is expressed in terms of the stresses or forces and variations of the complementary strain energy are affected by varying the stresses such that tractions vary only over those portions of the surface where the displacements are prescribed.

In the thermoelastic problem the above theorems are generalized to have the following forms. Thus the Theorem of Stationary Potential Energy becomes

$$\delta[\iiint F \mathrm{d}V - \iiint (Xu + Yv + Zw)\mathrm{d}V - \iint (X_s u + Y_s v + Z_s w)\mathrm{d}S] = 0 \qquad (1.44)$$

where, it is noted, the " free energy " F replaces the usual strain energy density. The Theorem of Stationary Complementary Potential Energy becomes

$$\delta[\iiint G_i \mathrm{d}V + \iint (uX_s + vY_s + wZ_s)\mathrm{d}S] = 0 \qquad (1.45)$$

where the surface integral is taken over that part of the surface where the displacements are prescribed, and the complementary strain energy density is replaced by the negative of the Gibbs Function $(-G_i)$.

It should be noted that Boley and Weiner[15] defined the strain energy per unit volume, U_o, by

$$U_o = \tfrac{1}{2}(e_{ij} - aT\delta_{ij})\sigma_{ij} \qquad (1.46)$$

(where δ_{ij} is the Kronecker delta) i.e. as the work done per unit volume by the internal stresses when they and the corresponding strains vary from zero to their terminal values. Similarly they defined the complementary strain energy per unit volume, U_o' by

$$U_o' = U_o + aT\Sigma \qquad (1.47)$$

When these equations are expressed solely in terms of strains alone and stresses alone, respectively, it is seen that U_o is equivalent to the " free energy " F per unit volume with $C_1(T')$ replaced by $\frac{3}{2}(3\lambda + 2\mu)(aT)^2$ and U_o' is equivalent to the negative of the Gibbs Function, G_i, per unit volume if $C_2(T')$ is zero.

The Theorem of Stationary Potential Energy becomes the Theorem of Stationary Strain Energy if the displacements are given over the entire surface and there are no body forces. The Theorem of Stationary Complementary Potential Energy becomes the Theorem of Stationary Complementary Strain Energy if the tractions are given over the entire surface (i.e. $W_u = 0$). This follows from the definition of W_u since in any problem the displacements and the tractions at a point on the surface cannot both be prescribed.

1.8. Solution of the Three-Dimensional Thermoelastic Equations

1.8.1. Various Formulations

If the temperature is given at all points in a solid body the stresses and displacements at the corresponding points can be determined by the solution of eqns. (1.9), (1.14) and (1.21)—if a rectangular coordinate system is used—subject to the appropriate boundary conditions of Section 1.5. This is, however, a very difficult problem except in the simplest cases since the following fifteen equations must be satisfied, viz.

6 strain–displacement relations (1.9)
6 stress–strain relations (1.14)
3 equilibrium equations (1.21)

in order to determine the following fifteen unknown functions

6 stress components: $\sigma_{xx} \ldots \sigma_{xy} \ldots$
6 strain components: $e_{xx} \ldots e_{xy} \ldots$
3 displacement components: u, v, w

Obviously if either a cylindrical or spherical coordinate system is used the appropriate equations of Sections 1.2 and 1.4 would replace eqns. (1.9) and (1.21), and the corresponding stress, strain and displacement components would be involved.

If the boundary conditions are given in terms of the tractions only (Section 1.5.1) or of the displacements only (Section 1.5.2) simplifications to the above formulation can be made by eliminating from the original fifteen equations those terms not entering directly into the boundary conditions. In this way either a stress formulation involving only six equations, or a displacement formulation involving only three equations, is obtained.

For the former, the six stress compatibility equations must be derived: e.g. from (1.12) by substitution. These, together with the equations of equilibrium, e.g. eqn. (1.21), and the boundary conditions (1.26) only contain the unknown six stress components. Two of these six stress compatibility equations are presented here, the remaining four can be written down with appropriate change of variable i.e.

$$\left.\begin{aligned}
&(1+\nu)\nabla^2\sigma_{xx} + \frac{\partial^2\Sigma}{\partial x^2} + E\left[\frac{1+\nu}{1-\nu}\nabla^2(\alpha T) + \frac{\partial^2(\alpha T)}{\partial x^2}\right] \\
&\quad + \nu\left(\frac{1+\nu}{1-\nu}\right)\left(\frac{\partial X}{\partial x} + \frac{\partial Y}{\partial y} + \frac{\partial Z}{\partial z}\right) + 2(1+\nu)\frac{\partial X}{\partial x} = 0 \\
&(1+\nu)\nabla^2\sigma_{xz} + \frac{\partial^2\Sigma}{\partial x\partial z} + E\frac{\partial^2(\alpha T)}{\partial x\partial z} \\
&\qquad\qquad + (1+\nu)\left(\frac{\partial X}{\partial z} + \frac{\partial Z}{\partial x}\right) = 0
\end{aligned}\right\} \tag{1.48}$$

For the latter formulation, the thermoelastic problem is reduced to solving the three equations of equilibrium—written in terms of the displacements—together with the appropriate boundary conditions (1.27), in order to obtain the displacement functions u, v and w throughout the body. A typical equation for this formulation is given here; the other two can be written down with appropriate change of variable, i.e.

$$(\lambda + \mu)\frac{\partial \bar{e}}{\partial x} + \mu\nabla^2 u - \beta\frac{\partial T}{\partial x} + X = 0 \tag{1.49}$$

where
$$\bar{e} = \frac{\partial u}{\partial x} + \frac{\partial v}{\partial y} + \frac{\partial w}{\partial z}$$

It should be noted that eqn. (1.48) applies directly to simply connected bodies but for multiply connected bodies further additional conditions must also be satisfied. No such restriction applies to eqn. (1.49) which holds for both simply and multiply connected bodies. This point is discussed further in Ref. 15 which also quotes in full the above sets of equations for both formulations in rectangular coordinates; and for the displacement formulation also in cylindrical and spherical coordinates.

In eqn. (1.48) generality has been maintained by retaining the thermal strain, αT, as a parameter. This may also be done in the displacement formulation.

1.8.2 Zero Displacements in a Three-Dimensional Body; and a Corollary

If a three-dimensional body is completely restrained in every direction ($e_{ij} = 0$) the equations of state (1.13) show that an equivalent " hydrostatic pressure ", p, must act such that

$$\sigma'_{xx} = \sigma'_{yy} = \sigma'_{zz} = p = -E\alpha T/(1-2\nu) \; ;$$
$$\sigma'_{xy} = \sigma'_{yz} = \sigma'_{zx} = 0 \tag{1.50}$$

Substitution of these expressions into the traction boundary conditions (1.26) indicates that the surface tractions necessary to maintain zero displacements are

$$X_s/l = Y_s/m = Z_s/n = -E\alpha T/(1-2\nu) = -\beta T \tag{1.51}$$

Similarly, the equations of equilibrium (1.21) indicate that the corresponding body forces for this situation are,

$$X = \beta\frac{\partial T}{\partial x} \; ; \; Y = \beta\frac{\partial T}{\partial y} \; ; \; Z = \beta\frac{\partial T}{\partial z} \tag{1.52}$$

In an unrestrained body the thermal stresses can be determined by adding to the stresses (1.50) those produced in the body by the force systems of (1.51) and (1.52) with reversed signs. These latter stresses are denoted by σ''_{ij} and satisfy the appropriate equations of equilibrium and the traction boundary conditions. Therefore, the solution of the general thermal stress problem in an unrestrained body may be considered as the sum of the two individual solutions above, i.e. $\sigma_{ij} = \sigma'_{ij} + \sigma''_{ij}$.

A similar result can be deduced from the displacement formulation, viz. that there exists an analogy (the " Body Force " analogy) between the thermoelastic problem and the isothermal problem for a body with distributed body forces. Equation (1.49) shows that the terms X and $-\beta(\partial T/\partial x)$ are analogous in that formulation. This analogy in no way reduces the mathematical complexity of a given problem but it may help to provide a clearer physical interpretation of a given thermoelastic problem, provided the material behaviour remains elastic.

1.8.3. Zero Stresses in a Three-Dimensional Body

If a structure which is free of body forces and surface tractions experiences a non-uniform temperature distribution then, in general, stresses and strains occur in the structure. If the stresses are all to be zero, the six stress compatibility equations (1.48) yield the results.

$$\frac{\partial^2}{\partial x^2}(\alpha T) = \frac{\partial^2}{\partial y^2}(\alpha T) = \frac{\partial^2}{\partial z^2}(\alpha T) = \frac{\partial^2 (\alpha T)}{\partial x \partial y} = \frac{\partial^2(\alpha T)}{\partial y \partial z} = \frac{\partial^2(\alpha T)}{\partial z \partial x} = 0 \qquad (1.53)$$

Therefore the thermal strain distribution must be a linear function of the *rectangular space coordinates* (x, y, z), i.e.

$$\alpha T = a_0 + a_1 x + a_2 y + a_3 z \qquad (1.54)$$

where the coefficients may be time dependent. For a material with constant thermal properties the temperature distribution is linear for zero stresses, although finite strains and displacements still occur.

It should be noted that if $T = Ar$, where r is the radial coordinate, in an unrestrained circular plate, there are thermal stresses since T is only linear in r and not in the rectangular coordinates, i.e. $r = (x^2 + y^2)^{1/2}$.

1.8.4. Summary of Methods of Solution

In addition to the " Body Force " analogy of Section 1.8.2, two other general methods of solution follow from the displacement formulation of Section 1.8.1. These are (a) Goodier's method[15, 55, 163] in which the thermoelastic problem is reduced to an isothermal problem with no body forces, and (*b*) the use of Boussinesq–Papkovich functions.[15,117,155] Unfortunately, all of these alternative formulations do not essentially simplify the solution of the more general three-dimensional problems and no further consideration of them will be given here. Only in isolated cases have such procedures found useful application, e.g. the study, by Goodier's method,[163] of thermal inclusions in solid bodies, and of hot-spots in thin plates (see Section 3.1).

Exact solutions may also be derived by the use of the energy methods of Section 1.7, which are readily adapted for approximate analyses based on strength of materials theory.

In many three-dimensional problems some simplification is possible by using Saint-Venant's principle[163] by which the rigorous traction boundary conditions of Section 1.5.1 are relaxed. If, also, the simple analytical procedures of the theory of strength of materials are used, many three-dimensional structural forms can be analyzed with sufficient accuracy, e.g. beams, rings or plates.

Numerical methods may be used to obtain solutions to thermal stress problems. Such methods may involve either influence coefficient or finite difference procedures. The former approach is especially useful for built-up structures of complex shape,[5,6] when elementary theories of strength of materials type may not be applicable. Among the most promising of these procedures are those in which the force–displacement relationships, for the various discrete elements comprising the structure, are cast in matrix form. Depending upon which quantities are left as unknowns in the analysis, these methods may be grouped into three general categories, viz. force methods, displacement methods or mixed force–displacement methods. Examples of each are given in papers by Denke[24], Turner *et al.*[167,168], and Klein[93]. The automatic formulation capabilities of these matrix methods, which are ideally suited for use with digital computation, are particularly advantageous when applied to, say, **non-linear** thermal stress problems.[46]

The use of the Airy stress function in two dimensional isothermal problems is well known[163]—its application for thermal stress analyses is discussed in Chapter 2.

CHAPTER 2

Two-Dimensional Formulations and Solutions

IT WAS shown in Chapter 1 that the general three-dimensional thermoelastic problem requires the determination of 15 quantities, viz. 6 stresses, 6 strains and 3 displacements, when the body forces and boundary conditions are known. The problem becomes much simpler if some of the unknown quantities are zero or insignificant as a result of the particular geometry or loading. Such is the case for *plane strain* and *plane stress* problems.

2.1. Plane Strain Analyses

The condition of plane strain arises when the displacement component in a given direction is zero and the other displacement components are independent of this direction. Therefore if w in the z-direction is zero, the conditions which define plane strain are

$$u = u_{(x,y)} \; ; \; v = v_{(x,y)} \; ; \; w = 0 \tag{2.1}$$

It can be shown that this condition occurs in a prismatic body whose length is large compared with its cross-sectional dimensions, and for temperatures and loads which are independent of the z coordinate; the body force component Z must also be zero.

With these restrictions the general three-dimensional theory is automatically satisfied, with

$$\sigma_{xx} = f_{1(x,y)}, \; \sigma_{yy} = f_{2(x,y)}, \; \sigma_{xy} = f_{3(x,y)}, \; \sigma_{xz} = \sigma_{yz} = 0 \tag{2.2}$$

and $$\sigma_{zz} = f_{4(x,y)}, \text{ when } T = T_{(x,y)}$$

In fact, the third equation (1.13) yields

$$\sigma_{zz} = \nu(\sigma_{xx} + \sigma_{yy}) - E\alpha T = f_{4(x,y)} \tag{2.3}$$

and this equation defines the tractions which are necessary on the end faces of the body to maintain the state of plane strain, i.e. $e_{zz} = 0$. Since these tractions are not, in general, equal to those required it is necessary to add to the plane strain solution, another solution which will make the end tractions have their required value. This secondary solution requires, in general, the analysis of a non-thermal, but three-dimensional problem. Thus if it is required that the surface tractions on the end faces should be zero

i.e. $$\sigma_{zz} = \sigma_{xz} = \sigma_{yz} = 0 \text{ on } z = 0,L \tag{2.4}$$

the secondary solution necessary must satisfy the conditions

$$\sigma_{xz} = \sigma_{yz} = 0 \text{ and } \sigma_{zz} = -f_{4(x,y)} \text{ on } z = 0,L \tag{2.5}$$

with the appropriate boundary conditions on the other faces. In specifying these conditions due note has been taken of the fact that in the plane strain solution the end shear stresses are already zero. In general, the secondary solution satisfying (2.5) is difficult to obtain, but by invoking Saint-Venant's principle a realistic but approximate solution is found for bodies whose length is much greater than their cross-sectional dimensions.

This principle, which enables modifications to be made to the boundary conditions of a given problem, states that, if the forces acting on a small portion of the surface of an elastic body are replaced by another statically equivalent set, only the local stress distribution is significantly altered; at other more distant points in the body the resultant error is negligible. The approximate secondary solution therefore, has the form

$$\left.\begin{aligned} \sigma_{zz} &= \frac{P}{A} + \frac{M_y \,.\, x}{I_{yy}} + \frac{M_x \,.\, y}{I_{xx}} \\ \sigma_{xy} = \sigma_{xx} &= \sigma_{yy} = \sigma_{xz} = \sigma_{yz} = 0 \end{aligned}\right\} \tag{2.6}$$

if it is assumed that the coordinate system is based on x and y axes which are both centroidal and principal, and, where,

$$\left.\begin{aligned} P &= -\iint f_{4(x,y)}\mathrm{d}x\mathrm{d}y & A &= \iint \mathrm{d}x\mathrm{d}y \\ M_y &= -\iint f_{4(x,y)} \,.\, x \,.\, \mathrm{d}x\mathrm{d}y & I_{yy} &= \iint x^2\mathrm{d}x\mathrm{d}y \\ M_x &= -\iint f_{4(x,y)} \,.\, y \,.\, \mathrm{d}x\mathrm{d}y & I_{xx} &= \iint y^2\mathrm{d}x\mathrm{d}y \end{aligned}\right\} \tag{2.7}$$

Therefore the required solution is found by summing eqns. (2.2) and (2.6)

2.2. Plane Stress Analyses

The condition of plane stress is defined as being a two-dimensional state of stress, e.g.

$$\sigma_{zz} = \sigma_{xz} = \sigma_{yz} = 0 \tag{2.8}$$

Substituting these values into the three-dimensional stress-compatibility equations, (1.48) with the body forces zero, leads to the conclusion that the temperature distribution must satisfy the equation,

$$\nabla^2(\alpha T) = F(z, t) \tag{2.9}$$

if a solution to the plane stress problem is also to satisfy exactly the three-dimensional theory. Therefore the assumptions of plane stress are less satisfactory than those for plane strain since they result in a more restrictive form of temperature distribution (cf. (2.9) with (2.2)) and only for this form of distribution can the assumption of plane stress be rigorously applied. One important class of problems satisfying eqn. (2.9) is that of a free plate with a temperature variation through the thickness only (Section 2.6).

In general however, for very thin prismatic bodies where the axial dimension is very small compared with the cross-sectional dimensions in the x, y plane, it would seem reasonable to suppose that the stresses in (2.8) should be small when the end faces are free of tractions, and that even if they are not exactly zero, this assumption would be a good approximation to reality.

An attempt to justify the plane stress hypothesis, for temperature distributions more general than those satisfying (2.9), is given in reference 15 starting from less restrictive assumptions than (2.8), viz.

$$\sigma_{zz} = \sigma_{xz} = \sigma_{yz} = 0 \text{ on } z = \pm \frac{d}{2} \tag{2.10}$$

It was assumed further that all the stress components as well as the temperature could be written in a power series in the coordinate z. Only the case of symmetry about the plane $z = 0$ was considered and the series adopted had the following forms,

$$T_{(x,y,z)} = \sum_{n=0}^{\infty} T^{(n)}_{(x,y)} z^{2n} \tag{2.11}$$

$$\sigma_{ij(x,y,z)} = \sum_{n=0}^{\infty} \sigma^{(n)}_{ij(x,y)} z^{2n} \text{ for } i,j = x,y \tag{2.12}$$

$$\sigma_{iz(x,y,z)} = \sum_{n=1}^{\infty} \sigma^{(n)}_{iz(x,y)} \left[z^{2n+1} - z \left(\frac{d}{2}\right)^{2n} \right] \text{for } i = x, y \tag{2.13}$$

$$\sigma_{zz(x,y,z)} = \sum_{n=2}^{\infty} \sigma^{(n)}_{zz(x,y)} \left[z^{2n} - nz^2 \left(\frac{d}{2}\right)^{2n-2} + (n-1) \left(\frac{d}{2}\right)^{2n} \right] \tag{2.14}$$

These expressions satisfy eqn. (2.10) and also the third equilibrium equation on $z = \pm d/2$ i.e.

$$\frac{\partial \sigma_{zz}}{\partial z} = 0 \text{ on } z = \pm \frac{d}{2} \tag{2.15}$$

These series are then substituted into the three equations of equilibrium (1.21) and the six stress compatibility equations (1.48). By examining terms in ascending power of z and d it can be shown that the plane stress solution obtained satisfies the three-dimensional thermoelastic equations up to terms of the lowest order in the thickness, i.e. $n = 0$. Thus for *thin* bodies temperature distributions of the form $T = T^{(0)}_{(x,y)}$ are admissible for plane stress solutions of the form of eqn. (2.8). Since linear variations in temperature do not cause thermal stress (Section 1.8.3) the distribution $T = (z/d)\, T_{(x,y)}$ is also admissible.

2.3. Summary of the Thermal Stress Equations in Two Dimensions

For two-dimensional problems of plane stress or plane strain the three-dimensional equations of thermoelasticity can be reduced in the following way.

The strain–stress relations (1.13) and the strain–displacement relations (1.9) become, for the case of plane stress in thin bodies

$$\left.\begin{aligned} e_{xx} &= \frac{1}{E}\left[\sigma_{xx} - \nu\sigma_{yy}\right] + \alpha T = \frac{\partial u}{\partial x} \\ e_{yy} &= \frac{1}{E}\left[\sigma_{yy} - \nu\sigma_{xx}\right] + \alpha T = \frac{\partial v}{\partial y} \\ e_{xy} &= \frac{1}{G}\sigma_{xy} = \frac{\partial u}{\partial y} + \frac{\partial v}{\partial x} \end{aligned}\right\} \quad (2.16)$$

These relations also apply to the problems of plane strain if the following substitutions are made, E_1, for E; ν_1 for ν; α_1 for α; where,

$$\left.\begin{aligned} E_1 &= E/(1 - \nu^2) \\ \nu_1 &= \nu/(1 - \nu) \\ \alpha_1 &= \alpha(1 + \nu) \end{aligned}\right\} \quad (2.17)$$

Note that $G = E/2(1 + \nu) = E_1/2(1 + \nu_1)$ and remains the same for both formulations.

The compatibility equations in terms of strains (1.12) reduce to,

$$\frac{\partial^2 e_{xx}}{\partial y^2} + \frac{\partial^2 e_{yy}}{\partial x^2} = \frac{\partial^2 e_{xy}}{\partial x \partial y} \quad (2.18)$$

and the equilibrium equations (1.21) become,

$$\left.\begin{aligned} \frac{\partial \sigma_{xx}}{\partial x} + \frac{\partial \sigma_{xy}}{\partial y} + X = 0 \\ \frac{\partial \sigma_{xy}}{\partial x} + \frac{\partial \sigma_{yy}}{\partial y} + Y = 0 \end{aligned}\right\} \quad (2.19)$$

Using the equilibrium equations, the compatibility condition in terms of stress components becomes,

$$\left(\frac{\partial^2}{\partial x^2} + \frac{\partial^2}{\partial y^2}\right)(\sigma_x + \sigma_y + E\alpha T) = -(1 + \nu)\left(\frac{\partial X}{\partial x} + \frac{\partial Y}{\partial y}\right) \quad (2.20)$$

where for plane strain the constants of eqn. (2.17) must be used.

Therefore the two-dimensional problem reduces to the determination of the stress and strain components which satisfy the compatibility equation, (2.20), the equilibrium equations (2.19) and the appropriate boundary conditions. For plane strain $e_{zz} = 0$ and hence

$$\sigma_{zz} = \nu(\sigma_{xx} + \sigma_{yy}) - E\alpha T \tag{2.21}$$

whilst for plane stress $\sigma_{zz} = 0$ and hence

$$e_{zz} = -\frac{\nu}{E}(\sigma_{xx} + \sigma_{yy}) + \alpha T \tag{2.22}$$

2.4. Use of the Airy Stress Function for Solid Structures

In the absence of body forces, $X = Y = 0$, eqns. (2.19) are satisfied if a stress function ϕ, known as the Airy stress function, is defined such that

$$\sigma_{xx} = \frac{\partial^2\phi}{\partial y^2}, \quad \sigma_{yy} = \frac{\partial^2\phi}{\partial x^2}, \quad \sigma_{xy} = -\frac{\partial^2\phi}{\partial x \partial y} \tag{2.23}$$

Equation (2.20) accordingly becomes,

$$\nabla^4\phi + E\alpha\nabla^2 T = 0 \tag{2.24}$$

where

$$\nabla^2 = \frac{\partial^2}{\partial x^2} + \frac{\partial^2}{\partial y^2} \tag{2.25}$$

When body forces are present, in addition to temperature, a potential (V) is assumed to exist such that

$$X = \frac{-\partial V}{\partial x}; \quad Y = \frac{-\partial V}{\partial y} \tag{2.26}$$

Equations (2.19) then become,

$$\left.\begin{aligned} \frac{\partial}{\partial x}(\sigma_{xx} - V) + \frac{\partial}{\partial y}(\sigma_{xy}) &= 0 \\ \frac{\partial}{\partial x}(\sigma_{xy}) + \frac{\partial}{\partial y}(\sigma_{yy} - V) &= 0 \end{aligned}\right\} \tag{2.27}$$

and these equations may be satisfied by taking

$$\sigma_{xx} - V = \frac{\partial^2\phi}{\partial y^2}, \quad \sigma_{yy} - V = \frac{\partial^2\phi}{\partial x^2}, \quad \sigma_{xy} = -\frac{\partial^2\phi}{\partial x \partial y} \tag{2.28}$$

in which ϕ is once again the stress function. Equation (2.20) then yields

$$\nabla^4\phi + E\alpha\nabla^2 T + (1 - \nu)\nabla^2 V = 0 \tag{2.29}$$

All the above formulae apply to the case of plane stress and the constants of (2.17) must be used in the case of plane strain.

The corresponding boundary conditions for traction free surfaces become[15]

$$\phi = \frac{\partial \phi}{\partial n} = 0 \tag{2.30}$$

where n is normal to the surface.

It follows from eqn. (2.23) that if there are zero stresses in the x, y plane ϕ must be a linear function of x and y which from (2.30) implies $\phi = 0$; hence, from (2.24) a necessary condition for zero stresses in the plane is,

$$\frac{\partial^2 T}{\partial x^2} + \frac{\partial^2 T}{\partial y^2} = \nabla^2 T = 0 \tag{2.31}$$

An identical result, also for zero body forces, is obtained directly from eqn. (2.20). Therefore, provided that the temperature distribution is plane harmonic and independent of z the stresses in the x, y plane are zero. The only non-zero stress component is σ_{zz} for the case of plane strain, which becomes,

$$\sigma_{zz} = -E\alpha T \tag{2.32}$$

For many problems in two-dimensional thermoelasticity it is preferable to use polar rather than rectangular cartesian coordinates. In this case the two appropriate equilibrium equations follow from eqn. (1.24) with terms in z neglected, and the stress function ϕ is related to the stress components by

$$\left.\begin{aligned} \sigma_{rr} &= \frac{1}{r}\frac{\partial \phi}{\partial r} + \frac{1}{r^2}\frac{\partial^2 \phi}{\partial \theta^2} \\ \sigma_{\theta\theta} &= \frac{\partial^2 \phi}{\partial r^2}, \quad \sigma_{r\theta} = -\frac{\partial}{\partial r}\left(\frac{1}{r}\frac{\partial \phi}{\partial \theta}\right) \end{aligned}\right\} \tag{2.33}$$

where
$$\left.\begin{aligned}\frac{\partial\sigma_{rr}}{\partial r}+\frac{1}{r}\frac{\partial\sigma_{r\theta}}{\partial\theta}+\frac{\sigma_{rr}-\sigma_{\theta\theta}}{r}+R=0\\ \frac{1}{r}\frac{\partial\sigma_{\theta\theta}}{\partial\theta}+\frac{\partial\sigma_{r\theta}}{\partial r}+\frac{2\sigma_{r\theta}}{r}+\Theta=0\end{aligned}\right\} \qquad (2.34)$$

2.5. One-Dimensional Thermal Stresses in a Thin Rectangular Slab

The problem of determining the correct thermal stress distributions for general two-dimensional temperature distributions in an unrestrained thin slab of rectangular planform is of considerable interest and many analyses on this topic have now been published. Some of these will be examined in Chapter 3, but for present purposes the special case of a one-dimensional temperature distribution will be considered here. The solution presented forms a convenient basis on which the more general two-dimensional solutions may be derived.

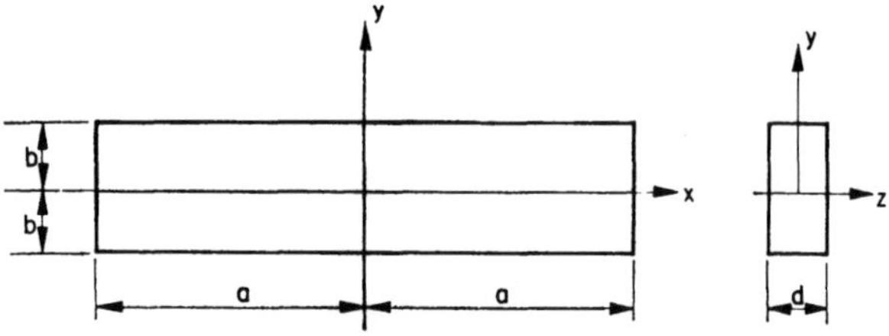

FIG. 2.1. Thin rectangular slab with a temperature distribution only through the depth.

Consider the thin rectangular slab of Fig. 2.1 in which the temperature variation is a function only of the y coordinate, and where the overall length $2a$ is very much greater than either the width $2b$ or the thickness d. For the thin slab with $T = T_{(y)}$ it can be assumed that the plane stress assumption holds, viz.

$$\sigma_{zz} = \sigma_{xz} = \sigma_{yz} = 0 \qquad (2.35)$$

If it is also assumed that the body forces and the stresses σ_{yy} and σ_{xy} are zero and $\sigma_{xx} = f_{(y)}$ the equilibrium equations (2.19) are satisfied and the compatibility equation (2.20) reduces to

$$\frac{d^2}{dy^2}(\sigma_{xx} + E\alpha T) = 0 \tag{2.36}$$

The solution for σ_{xx} is therefore,

$$\sigma_{xx} = -E\alpha T + C_1 + C_2 y \tag{2.37}$$

from which it follows that the tractions at the ends of the slab ($x = \pm a$) cannot be zero at all points across the width $2b$, except for the case when $T_{(y)}$ is linear in y, when all the stress components would necessarily be zero. Therefore, if the condition of zero resultant force and moment on the yz plane is invoked the constants C_1, C_2 may be determined, producing,[163]

$$\sigma_{xx} = -E\alpha T + \frac{1}{2b}\int_{-b}^{b} E\alpha T \, dy + \frac{3y}{2b^3}\int_{-b}^{b} E\alpha T y \, dy \tag{2.38}$$

The similarity between this result and that of Section 2.1 should be noted. In fact, eqn. (2.38) could as easily have been obtained by treating the present problem as one of plane strain, but with only one non-zero stress component, $\sigma_{xx(y)} = -E\alpha T_{(y)}$. The analysis would then have followed that of Section 2.1 exactly to give eqn. (2.38).

It should be emphasized that the above plane stress solution satisfies the general, three-dimensional, thermoelastic equations rigorously, apart from local effects near to the ends of the slab, i.e. within the approximation implicit in applying Saint-Venant's principle. The above solution may relate to a plate with dimension d corresponding to its thickness, or to a beam with $2b$ corresponding to its thickness (and d being its width).

2.6. Thick Plate with Temperature Variation through the Thickness Only

For the thick plate shown in Fig. 2.2, the various stress components, assuming all body forces and surface tractions are zero, may be considered to be

$$\sigma_{xx} = \sigma_{yy} = F_{(z)}, \; \sigma_{zz} = \sigma_{xz} = \sigma_{yz} = \sigma_{xy} = 0, \quad \text{when } T = T_{(z)} \tag{2.39}$$

These assumptions enable the three-dimensional equations of Chapter 1 to be satisfied exactly provided that,[15] from (1.48),

$$\frac{d^2}{dz^2}\left\{F + \frac{E\alpha T}{1-\nu}\right\} = 0 \tag{2.40}$$

The similarity of this result with that of the previous section should be noted, from which it follows that the required solution is

$$\sigma_{xx} = \sigma_{yy} = \frac{1}{1-\nu}\left[-E\alpha T + \frac{1}{d}\int_{-d/2}^{d/2} E\alpha T\,dz + \frac{12z}{d^3}\int_{-d/2}^{d/2} E\alpha Tz\,dz\right] \tag{2.41}$$

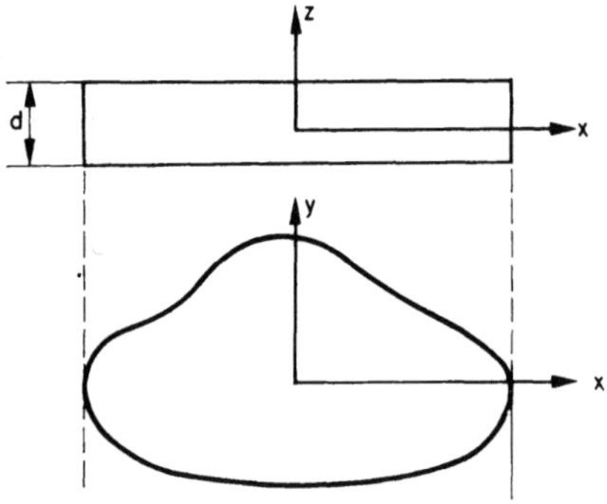

FIG. 2.2. Thick plate with a temperature distribution only through the thickness.

Again, this solution is quite rigorous apart from local effects near to the plate edges.

The term $-E\alpha T/(1-\nu)$ corresponds to complete suppression of the thermal expansions in the plane of the plate whilst the other two terms, which apply to a free plate, satisfy the conditions of zero net force and moment on the plate.[163]

Equation (2.41) may be rewritten as

$$\sigma_{xx} = \sigma_{yy} = \frac{E\alpha}{1-\nu}\left[T_m + \Delta T\frac{z}{d} - T\right] \tag{2.42}$$

$$= \frac{-E\alpha T^*}{1-\nu} \tag{2.43}$$

where the temperature distribution $T_{(z,\,t)}$ has been separated into linear and self-equilibrating distributions, viz.

$$T_{(z,\,t)} = T_m + \Delta T \frac{z}{d} + T^* \tag{2.44}$$

where T_m is the mean temperature and ΔT is the linear gradient across the thickness, and T^* is the self-equilibrating distribution satisfying the equations

$$\int_{-d/2}^{d/2} T\mathrm{d}z = \int_{-d/2}^{d/2} Tz\mathrm{d}z = 0 \tag{2.45}$$

The mean temperature T_m induces stresses only if there is restraint against expansion in the plane of the plate and since these are equal to $E\alpha T_m/(1 - \nu)$ they may be very large in practice.

The temperature gradient ΔT produces a uniform curvature in the plate and, in the absence of edge restraints, the plane plate deforms into a spherical surface free of stress with radius of curvature $R = d/\alpha\Delta T$.

The stresses corresponding to T^* exist even with no edge restraints, i.e. eqns. (2.41) and (2.43). No general comments can be made on their magnitude.

The effect of various edge restraint conditions will be considered in more detail in Chapter 4.

2.7. Thermal Stresses in Thin Plates

A plate is a flexible three-dimensional structure which has one dimension (the thickness) much smaller than the other two dimensions. If the deflexions normal to the median plane of the plate are small compared with the thickness a very satisfactory *small deflexion* theory for thin plates can be developed in which it is assumed that there is negligible interaction between the membrane and bending actions of the plate. From the thermal stress point of view, i.e. for general two-dimensional temperature distributions typical plate problems may be divided into two categories:

(*a*) Membrane problems, which arise when the plate is loaded and/or heated uniformly through the thickness. That is, the temperature varies with the in-plane coordinates only, the problem is one of plane stress applicable to thin plates, and the analyses follow directly from the discussions of Sections 2.2–2.4.[143]

(*b*) Bending problems, which arise as a result of temperature variations through the thickness. This latter class of problems can be solved by techniques already well known in isothermal thin plate theory. Analyses in this category consider one of two classes of problem: those with temperature variations through the thickness only, i.e. $T = T_{(z)}$ as in Section 2.6; and those for which the temperature is only linear in z, i.e. $T = (z/d)\,T_{(x,\,y)}$.

The above two categories will be examined separately in the following two chapters—Chapters 3 and 4.

When the normal deflexions of the plate are not small compared with the plate thickness there is a coupling between membrane and bending actions of the plate and to solve such problems a *large deflexion* theory is required. This theory will be presented also, for convenience, in Chapter 4 but since its application is more usual in post-buckling analyses discussion on this topic will be left until Chapter 7 which deals with Thermal Buckling.

CHAPTER 3

Membrane Thermal Stresses in Thin Plates

THE analyses to be presented in this chapter are all based on the hypothesis of plane stress discussed in Chapter 2. However, as was emphasized in Section 2.3, results obtained from plane stress analyses may be applied directly to problems of plane strain by substituting the appropriate parameters of eqn. (2.17).

3.1. Thermal Stresses Near Hot Spots in Infinite Plates

An interesting method of solution to this two-dimensional problem is given by Goodier[55,163] as a special case of a more general three-dimensional analysis based on the displacement formulation of Section 1.8.1. It can be shown that the three-dimensional displacement compatibility equations (1.49) are all identically satisfied if a *potential function* $\psi_{(x,y,z)}$ can be found, from which the displacement components may be written as follows, i.e.

$$u = \frac{\partial \psi}{\partial x}, \quad v = \frac{\partial \psi}{\partial y}, \quad w = \frac{\partial \psi}{\partial z} \tag{3.1}$$

and where the function ψ is a solution of the equation

$$\nabla^2 \psi = \frac{1+\nu}{1-\nu} \alpha T \tag{3.2}$$

This equation has been derived assuming zero body forces, and leads to the following forms of equation in two-dimensional problems. For problems of plane strain, when $w = 0$ and T is independent of z, (3.2) becomes

$$\frac{\partial^2\psi}{\partial x^2} + \frac{\partial^2\psi}{\partial y^2} = \frac{1+\nu}{1-\nu}\,\alpha T \tag{3.3}$$

with ψ, u and v all independent of z; and the corresponding plane stress equation is

$$\frac{\partial^2\psi}{\partial x^2} + \frac{\partial^2\psi}{\partial y^2} = (1+\nu)\,\alpha T \tag{3.4}$$

A particular solution of (3.4) is given by the *logarithmic potential*,

$$\psi = \frac{(1+\nu)}{2\pi}\alpha \iint T_{(\xi,\,\eta)} \log r' \mathrm{d}\xi \mathrm{d}\eta \tag{3.5}$$

where $T_{(\xi,\,\eta)}$ is the temperature at a typical point ξ, η and r' is the distance between this point and the point x, y, i.e.

$$r' = [(x-\xi)^2 + (y-\eta)^2]^{1/2} \tag{3.6}$$

Equation (3.5) gives the full solution for localized heating in an infinite plate where the deformation and stress must tend to zero at infinity.

For a *rectangular hot spot*[55] of sides $2a$, $2b$ and at temperature T in an infinite plate at zero temperature, the required logarithmic potential is

$$\psi = \frac{(1+\nu)}{2\pi}\,\alpha T \int_{-b}^{b}\int_{-a}^{a} \tfrac{1}{2}\log\Big[\,(x-\xi)^2 + (y-\eta)^2\,\Big]\mathrm{d}\xi\mathrm{d}\eta \tag{3.7}$$

By differentiating this result the displacements are found, and by using the stress–strain relations for plane stress problems expressions are obtained for the stresses σ_{xx}, σ_{yy} and σ_{xy}. The maximum value of the direct stresses varies from $0{\cdot}375E\alpha T$ for a *square hot spot* to $0{\cdot}5E\alpha T$ for a hot spot of infinite length. As a consequence of assuming an ideally sharp corner to the heated rectangle the shear stress σ_{xy}, approaches infinity as a corner is approached.

For an *elliptical hot spot* defined by

$$\left(\frac{x^2}{a^2} + \frac{y^2}{b^2}\right) = 1 \tag{3.8}$$

the maximum direct stress varies from $0{\cdot}5E\alpha T$ for a *circular hot spot* to $E\alpha T$ for a very elongated elliptical hot spot ($b = 0$). In fact, the value of σ_{yy} just outside the ellipse and near the ends of the major axis ($x = a$) is

$$\sigma_{yy} = E\alpha T/\left(1 + \frac{b}{a}\right) \tag{3.9}$$

For the *circular hot spot* of radius a in an infinite plate, the complete expressions for the stresses are

$$\sigma_{rr} = \sigma_{\theta\theta} = -\,0{\cdot}5E\alpha T \text{ inside the hot spot} \tag{3.10}$$

and $$\sigma_{rr} = -\,\sigma_{\theta\theta} = -\,0{\cdot}5E\alpha Ta^2/r^2 \text{ outside the hot spot} \tag{3.11}$$

It should be mentioned that problems of three-dimensional hot inclusions can also be analyzed by means of eqn. (3.2). For *ellipsoidal inclusions* and similar problems references 122 and 146 should be consulted. For the ellipsoid the maximum direct stress varies from $\frac{2}{3}E\alpha T/(1 - \nu)$ for a *spherical inclusion* to $E\alpha T/(1 - \nu)$ for a very elongated ellipsoid. In the former case the complete expressions for the stresses are[163]

$$\sigma_{rr} = -\,2\sigma_{\theta\theta} = -\,2\sigma_{\phi\phi} = -\frac{2}{3}\frac{E\alpha T}{1 - \nu}\frac{a^3}{r^3} \text{ outside the heated region.} \tag{3.12}$$

3.2. A Circular Disc with a General Radial Temperature Distribution

In the absence of body forces and because of the axial symmetry of the temperature distribution, the equations of equilibrium in two-dimensional thermoelasticity (2.34) reduce to

$$\frac{\mathrm{d}\sigma_{rr}}{\mathrm{d}r} + \frac{1}{r}(\sigma_{rr} - \sigma_{\theta\theta}) = 0 \tag{3.13}$$

For this problem the strain–displacement and strain–stress relations are,

$$e_{rr} = \frac{\mathrm{d}u}{\mathrm{d}r} = \frac{1}{E}\left[\sigma_{rr} - \nu\sigma_{\theta\theta}\right] + \alpha T \tag{3.14}$$

$$e_{\theta\theta} = \frac{u}{r} = \frac{1}{E}\left[\sigma_{\theta\theta} - \nu\sigma_{rr}\right] + \alpha T \tag{3.15}$$

These equations give the following stress–strain relations,

$$\sigma_{rr} = \frac{E}{1 - \nu^2}\left[e_{rr} + \nu e_{\theta\theta} - (1 + \nu)\alpha T\right] \tag{3.16}$$

$$\sigma_{\theta\theta} = \frac{E}{1 - \nu^2}\left[e_{\theta\theta} + \nu e_{rr} - (1 + \nu)\alpha T\right] \tag{3.17}$$

Substituting these expressions into (3.13) and writing the equation in terms of the radial displacement, u, yields eventually

$$\frac{\mathrm{d}}{\mathrm{d}r}\left[\frac{1}{r}\frac{\mathrm{d}(ur)}{\mathrm{d}r}\right] = (1 + \nu)\alpha\frac{\mathrm{d}T}{\mathrm{d}r} \tag{3.18}$$

By integrating eqn. (3.18) the displacement function, u, may be found, and by eqns. (3.14) to (3.17) the stress components may be determined. Thus

$$u = \frac{(1 + \nu)\alpha}{r}\int_a^r Tr\,\mathrm{d}r + C_1 r + C_2/r \tag{3.19}$$

$$\sigma_{rr} = \frac{-E\alpha}{r^2}\int_a^r Tr\,\mathrm{d}r + EC_1/(1 - \nu) - EC_2/r^2(1 + \nu) \tag{3.20}$$

$$\sigma_{\theta\theta} = \frac{E\alpha}{r^2}\int_a^r Tr\,\mathrm{d}r - E\alpha T + EC_1/(1 - \nu) + EC_2/r^2(1 + \nu) \tag{3.21}$$

The appropriate boundary conditions of traction-free edges on a disc with a concentric hole enable the constants C_1 and C_2 to be determined, i.e. for $\sigma_{rr} = 0$, at $r = a$ and $r = b$, the inner and outer radii respectively. The resulting expressions are,

$$u = \frac{\alpha}{r}\left\{(1 + \nu)\int_a^r Tr\mathrm{d}r + \frac{(1 - \nu)r^2 + (1 + \nu)a^2}{b^2 - a^2}\int_a^b Tr\,\mathrm{d}r\right\} \tag{3.22}$$

$$\sigma_{rr} = \frac{E\alpha}{r^2}\left[\frac{r^2 - a^2}{b^2 - a^2}\int_a^b Tr\,\mathrm{d}r - \int_a^r Tr\,\mathrm{d}r\right] \tag{3.23}$$

$$\sigma_{\theta\theta} = \frac{E\alpha}{r^2}\left[\frac{r^2 + a^2}{b^2 - a^2}\int_a^b Tr\,\mathrm{d}r + \int_a^r Tr\,\mathrm{d}r - Tr^2\right] \tag{3.24}$$

In these equations it is assumed that the disc is free, i.e. $\sigma_{rr} = 0$ at $r = a$, $r = b$. If, however, the disc is now considered to be fully restrained on the outer circumferential edge, i.e. $u = 0$ at $r = b$, the additional problem of the radial restraint forces may be solved as an ordinary problem of isothermal elasticity. Conversely the original equations (3.19) and (3.20) may be solved directly using the appropriate boundary conditions, i.e. $\sigma_{rr} = 0$ at $r = a$; $u = 0$ at $r = b$. For elastic restraint on the outer edge, $r = b$, a relationship is assumed to exist between the quantities u and σ_{rr} of the form $U = \bar{K}\,\sigma_{rr}$ at $r = b$, where $\bar{K}$ is a flexibility coefficient. This relationship together with the condition $\sigma_{rr} = 0$ at $r = a$ enables the complete problem to be solved.

For a solid circular disc the corresponding boundary conditions are $u = 0$ at $r = 0$ and $\sigma_{rr} = 0$ at $r = b$. Thus $C_2 = 0$ from eqn. (3.19) and C_1 is determined from eqn. (3.20). The results for this case could also have been found from those above, i.e. eqns. (3.22)–(3.24), by setting $a = 0$. When this is done the expressions are found to be indeterminate for $r = 0$. Assuming that the temperature is finite at $r = 0$ the following limits are obtained for the " indeterminate " integrals, viz.

$$\left.\begin{aligned} \frac{1}{r^2}\int_0^r Tr\,\mathrm{d}r &= \tfrac{1}{2}\,T_{(0)} \quad \text{as } r \to 0 \\ \frac{1}{r}\int_0^r Tr\,\mathrm{d}r &= 0 \qquad\quad \text{as } r \to 0 \end{aligned}\right\} \tag{3.25}$$

In this case σ_{rr} is a maximum and equal to $\sigma_{\theta\theta}$ at $r = 0$, i.e.

$$\sigma_{rr(0)} = \sigma_{\theta\theta(0)} = E\alpha\left[\frac{1}{b^2}\int_0^b Tr\,\mathrm{d}r - \tfrac{1}{2}T_{(0)}\right] \tag{3.26}$$

There are many other analyses in the literature relating to axisymmetric thermal stresses, particularly for specific forms of temperature distributions[48] and boundary conditions.[159] Thus, in references 38 and 133 the problem of a circular disc with a central hot spot is considered. The steady state temperature distribution is assumed, i.e. $T = T_a \log(b/r)/\log(b/a)$, where b is the outer radius of the disc and a is the radius of the hot spot which is at a uniform temperature T_a above the edge of the disc. Both papers contain results for various special cases.

3.3. A Solid Circular Disc with an Asymmetrical Temperature Distribution

For this plane stress problem in a circular disc, Horvay[72] and Forray[39] assume that the quantity $E\alpha T$ in eqn. (2.24) may be written in the form of the series

$$- P_{0(r)} - \sum_{n=1}^{\infty} [P_{n(r)} \cos n\theta + Q_{n(r)} \sin n\theta] \text{ where}$$

$P_{n(r)}$ and $Q_{n(r)}$ are assumed to be known.

Since a particular solution, ϕ_p, of

$$\nabla^2\phi = - E\alpha T \tag{3.27}$$

is also a particular solution of eqn. (2.24) and since the complete solution of $\nabla^4\phi = 0$, ϕ_c, is already known,[163] the required solution of (2.24) may be written,

$$\phi = \phi_p + \phi_c \tag{3.28}$$

It is readily proved that,

$$\phi_p = g_0(r) + \sum_{n=1}^{\infty} [g_n(r) \cos n\theta + h_n(r) \sin n\theta] \tag{3.29}$$

where

$$g_n = r^{-n} \int r^{2n-1} \int P_{n(r)} r^{1-n} \mathrm{d}r \, \mathrm{d}r \; (n = 0, 1, 2, \ldots) \tag{3.30}$$

$$h_n = r^{-n} \int r^{2n-1} \int Q_{n(r)} r^{1-n} \mathrm{d}r \, \mathrm{d}r \; (n = 1, 2, \ldots) \tag{3.31}$$

and

$$\phi_c = C_{00} + C_{11}r^2 + [C_{12}r + C_{13}r^3] \cos\theta + [C'_{12}r + C'_{13}r^3] \sin\theta$$
$$+ \sum_{n=2}^{\infty} [(C_n r^n + d_n r^{n+2}) \cos n\theta + (C'_n r^n + d'_n r^{n+2}) \sin n\theta] \tag{3.32}$$

In writing (3.32) allowance has been made for the fact that the stresses must remain finite in a solid disc as $r \to 0$; certain terms, therefore, in the more general solution[163] have been omitted.

The coefficients in (3.32) are determined by satisfying the various boundary conditions, at the outer periphery of the disc, on the form of ϕ in (3.28), e.g. for traction-free surfaces $\phi = \partial\phi/\partial r = 0$ at $r = b$. Utilizing the above procedure the following results are obtained.[159] If $T = T_0(r/b)^K \cos n\theta$ ($K = 0, 1, 2 \ldots, n = 0, 1, 2, \ldots$) the formulae for the stress components are:

Case I, $K \neq n - 2$,

$$\left.\begin{aligned} \sigma_{rr} &= [\bar{a}\rho^K + \bar{b}\rho^{n-2} + c\rho^n]\, E\alpha T_0 \cos n\theta \\ \sigma_{\theta\theta} &= [A\rho^K + B\rho^{n-2} + C\rho^n]\, E\alpha T_0 \cos n\theta \\ \sigma_{r\theta} &= [\delta\rho^K + \beta\rho^{n-2} + \gamma\rho^n]\, E\alpha T_0 \sin n\theta \end{aligned}\right\} \quad (3.33)$$

Case II, $K = n - 2$,

$$\left.\begin{aligned} \sigma_{rr} &= [a'\rho^K + b'\rho^{n-2} ln.\rho + c\rho^n]\, E\alpha T_0 \cos n\theta \\ \sigma_{\theta\theta} &= [A'\rho^K + B'\beta^{n-2} ln.\rho + C\rho^n]\, E\alpha T_0 \cos n\theta \\ \sigma_{r\theta} &= [\delta'\rho^K + \beta'\rho^{n-2} ln.\rho + \gamma\rho^n]\, E\alpha T_0 \sin n\theta \end{aligned}\right\} \quad (3.34)$$

where $\rho = r/b$; $\bar{a} = (n^2 - K - 2)/L$; $A = -(K + 1)(K + 2)/L$;

$\delta = -n(K + 1)/L$;

$B = \beta = -\bar{b} = n(n-1)(n-K)/2L$;

$B' = \beta' = -b' = (n-1)/2$;

$c = (n + 1)(n - 2)/M$; $C = (n + 1)(n + 2)/M$;

$\gamma = n(n + 1)/M$;

$a' = (n + 1)(n - 2)/4n$; $A' = (2 - 3n - n^2)/4n$;

$\delta' = -(n + 1)/4$;

$L = (K + 2)^2 - n^2$; $M = 2(K + 2 + n)$.

It should be noted that since the results quoted are obtained for a linear elastic problem, the principle of superposition may be used if there are several terms in the expression for the temperature distribution. In reference 39 specific results are presented for the case $K = 0, 2$ and $n = 0, 1$.

Further, if $T = T_0(r/b)^K \sin n\theta$ ($n = 1, 2, \ldots$) then the results of (3.33) and (3.34) may be used directly if the terms $\cos n\theta$ and $\sin n\theta$ are replaced by $\sin n\theta$ and $-\cos n\theta$ respectively.

When $K = n$ the assumed temperature distribution satisfies the equation $\nabla^2 T = 0$, and it follows from (2.31) that the stresses in the disc are zero. When $n = 0$ there is radial symmetry, i.e. $T = T_0 (r/b)^K$, and the above results still apply.

3.4. A Circular Ring with an Asymmetrical Temperature Distribution

The method of analysis for this problem[40,41] follows very closely that outlined above for the solid circular disc. The essential difference is in the boundary conditions to be satisfied, viz. $\sigma_{rr} = \sigma_{r\theta} = 0$ at $r = a$ and $r = b$, the inner and outer radii of the ring. The form of temperature distribution is assumed to be $T = T_0(r/b)^K \cos n\theta$.

Specific results are presented in reference 40 for the case $K = 0, 2$, $n = 0, 1$; and in reference 159 for a particular radial power law temperature distribution, viz. $T = T_0[(r-a)/(b-a)]^K$. It should be noted that thermal stresses in rings of arbitrary cross-section and contour are considered in reference 15. The principle of stationary complementary strain energy is used to determine the unknown redundants of the heated ring in the absence of external loads, and it is assumed that the strain energy can be expressed simply in terms of the circumferential stress component, $\sigma_{\theta\theta}$, and the temperature. The assumption can be justified if the temperature distribution is independent of θ and the ring has a small depth to radius ratio.

3.5. Thermal Stresses in Finite Rectangular Plates (Method 1)

In Section 2.5 an analysis is presented for the stresses in a rectangular slab (Fig. 2.1) subjected to a one-dimensional temperature distribution. As applied to a thin plate, the temperature is assumed to vary across the width of the plate and if the temperature variation is taken to be

$$T_{(y)} = T_0 - \Delta T \left|\frac{y}{b}\right|^n \qquad n = 1, 2, 3, \ldots, \qquad (3.35)$$

the corresponding symmetrical thermal stress distribution is from (2.38),

$$\sigma_{xx} = -E\alpha\Delta T\left[\frac{1}{n+1} - \left|\frac{y}{b}\right|^n\right], \sigma_{yy} = \sigma_{xy} = 0 \qquad (3.36)$$

In deriving eqn. (2.38) and hence (3.36) Saint-Venant's principle has been invoked, in other words pointwise satisfaction of the traction-free boundary condition on the end faces $x = \pm a$ is not ensured, and the expressions (2.38) and (3.36) only apply to points in the plate remote from the end faces. For this reason such expressions are often termed the *infinite plate thermal stresses.*[143]

Because eqns. (2.38) and (3.36) give a distributed normal stress distribution on the end faces, the condition of a traction-free end face can be achieved by applying at the ends a stress distribution equal in magnitude but opposite in sign to the " infinite plate " stresses. This latter problem can be treated as one in isothermal elasticity[163] and the resulting two-dimensional stress distribution when superimposed on the original one-dimensional " infinite plate " stress distribution becomes the required correct stress distribution in the finite plate. It is found that this correct stress distribution contains significant shear stress terms near the ends of the plate (such stresses are zero in the infinite plate). Thus, the departure of the correct distribution from the infinite plate expression is known as the *shear lag effect.*

If the stress distribution $-\sigma_{xx}$ is applied to the end faces of the plate, then, from (3.36) the boundary conditions to be satisfied on the various edges of the plate are

$$\left.\begin{aligned} &\sigma_{xx} = E\alpha\Delta T\left[\frac{1}{n+1} - \left|\frac{y}{b}\right|^n\right], \ \sigma_{xy} = 0 \text{ at } x = \pm a \\ &\sigma_{yy} = \sigma_{xy} = 0 \qquad\qquad\qquad\qquad \text{at } y = \pm b \end{aligned}\right\} \qquad (3.37)$$

The complementary strain energy for a plate of unit thickness is given in this case by

$$\bar{U}_0' = \frac{1}{2E}\int_{-a}^{a}\int_{-b}^{b}\left[\left(\frac{\partial^2\phi}{\partial x^2}\right)^2 + \left(\frac{\partial^2\phi}{\partial y^2}\right)^2 + 2\left(\frac{\partial^2\phi}{\partial x\partial y}\right)^2\right]\mathrm{d}x\mathrm{d}y \qquad (3.38)$$

where ϕ is the Airy stress function. The absence of Poisson's ratio in this expression is admissible since, from (2.20), for an isothermal problem with constant body forces, the stress distribution is independent of the material elastic constants.

The correct expression for ϕ is that which satisfies the boundary conditions and makes the strain energy a minimum. For the symmetrical problem considered a suitable form for ϕ is[163]

$$\left.\begin{aligned}\phi &= \frac{E\alpha\Delta T b^2}{2(n+1)(n+2)}\left[(n+2)\left(\frac{y}{b}\right)^2 - 2\left|\frac{y}{b}\right|^{n+2}\right]\\ &\quad + (x^2-a^2)^2(y^2-b^2)^2[\mu_1+\mu_2x^2+\mu_3y^2+\ldots]\end{aligned}\right\} \quad (3.39)$$

Substituting this expression into eqn. (3.38) gives $\overline{U}_0'$ as a function of the constants μ_1, μ_2, μ_3, etc. The minimum conditions

$$\frac{\partial \overline{U}_0'}{\partial \mu_1} = \frac{\partial \overline{U}_0'}{\partial \mu_2} = \frac{\partial \overline{U}_0'}{\partial \mu_3} = \ldots = \frac{\partial \overline{U}_0'}{\partial \mu_m} = 0 \quad (3.40)$$

lead to m simultaneous equations from which each value of μ_m may be determined.

If μ_1 only is considered the minimization process yields

$$\mu_1 = \frac{105}{128}\,\frac{n}{(n+1)(n+3)}\,\frac{E\alpha\Delta T}{b^2(a^4+b^4+\frac{4}{7}a^2b^2)} \quad (3.41)$$

from which, ϕ, eqn. (3.39), and then the stresses, eqn. (2.23), may be determined. These stresses when added to those of eqn. (3.36) give the required correct thermal stress distribution.

Obviously better results would be obtained by taking more terms μ_m into the analysis. Thus, Fig. 3.1 presents corresponding results for both a square plate ($a = b$) and a rectangular plate ($a = 2b$) with $n = 2$; and from analyses using (i) only μ_1 and (ii) the three parameters μ_1, μ_2 and μ_3. The stress distributions shown are for σ_{xx} at the centre of the plate $x = 0$, and as the " infinite plate " thermal stresses are shown for comparison it may be seen that:

(1) The three-term solution is only slightly better than the one-term solution.

(2) In the rectangular plate there is very little difference between the " correct " and the " infinite plate " stresses at $x = 0$, whereas in the square plate there is a significant difference. This indicates that " shear-lag " effects are insignificant at distances from the ends greater than the maximum cross-sectional dimension (i.e. the width $2b$) and gives particular meaning to Saint-Venant's principle.

3.6. Thermal Stresses in Finite Rectangular Plates (Methods 2–4)

The analysis above can only be applied for a one-dimensional temperature distribution and alternative procedures are necessary for two-dimensional temperature distributions i.e. $T = T_{(x, y)}$.

Mendelson and Hirschberg[115] (Method 2) use a collocation procedure whereby the partial differential equation for the Airy stress function (2.24) is satisfied everywhere in x but at only a finite

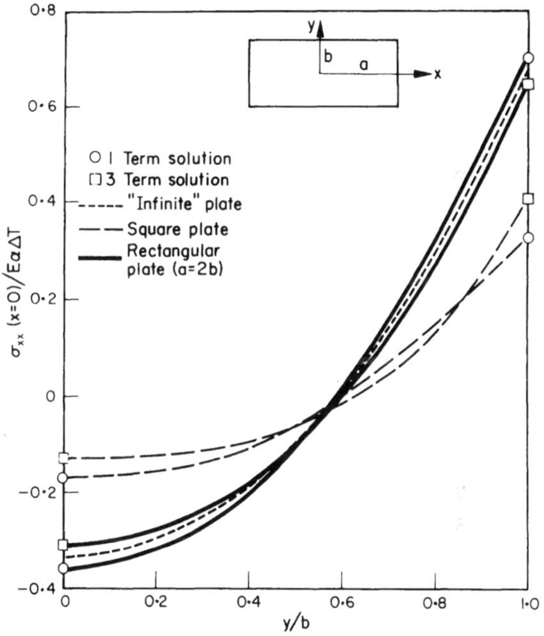

FIG. 3.1. Thermal stress at centre of rectangular plate for a parabolic temperature distribution (Method 1)—Taken from reference 143. (*J. Aero. Soc. of India*).

number of values of "y". The problem is thus reduced to the solution of a system of simultaneous fourth-order ordinary differential equations for some unknown functions of y. The main disadvantage of the collocation method is the considerable amount of computation necessary, but with high speed computing equipment this problem is readily overcome. In reference 144 an extensive set of graphical and tabular results is presented enabling solutions to be found for a wide range of problems for plate aspect ratios (b/a) between 0·1 and 3·0.

Przemieniecki[141] (Method 3) presents a method of solution derived directly from eqn. (2.24) in which the Airy stress function is expressed as a generalized Fourier expansion in terms of the

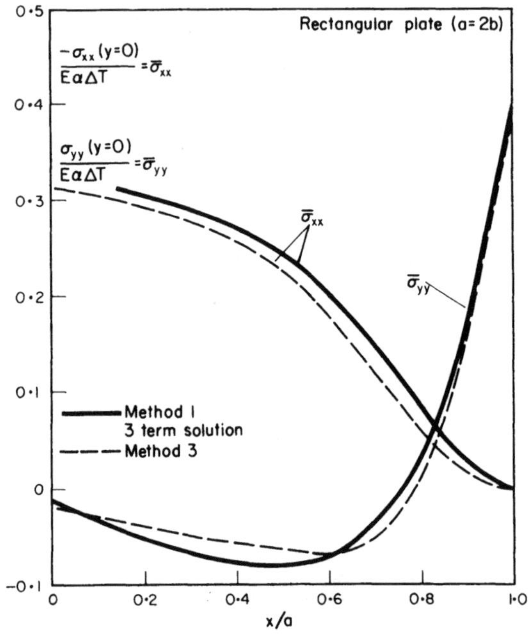

FIG. 3.2. Thermal stresses in a rectangular plate for a parabolic temperature distribution (Methods 1 and 3)—Taken from reference 143. (*J. Aero. Soc. of India*).

characteristic functions (eigenfunctions) representing normal modes of vibration of a uniform clamped–clamped beam. In this way the boundary conditions are satisfied exactly, but eqn. (2.24) is only approximately satisfied, depending upon the number of the terms in the Fourier expansion. Results obtained by this method and Method 1 are shown in Fig. 3.2, for a rectangular plate $(a/b = 2)$ and with $n = 2$, i.e. for a parabolic one-dimensional temperature distribution. Figure 3.2 illustrates well the manner in which the longitudinal stress σ_{xx} (at $y = 0$) decreases from its maximum value at

$x = 0$ to zero at the end of the plate, $x = a$. It is also seen that the transverse stress σ_{yy} (at $y = 0$), introduced by the "shear-lag" effect near the end of the plate is of the same order of magnitude as the stress σ_{xx}.

Heldenfels and Roberts[62] (Method 4) use an approximate variational method in which the Airy stress function is assumed to be given by,

$$\phi = f_{(x)}\, g_{(y)} \tag{3.42}$$

and the temperature distribution by

$$T_{(x,\, y)} = T_0 + X_{(x)}\, Y_{(y)} \tag{3.43}$$

It is further assumed that the function g is given by the one-dimensional solution for an infinite plate with $T = Y_{(y)}$. Therefore from eqns. (2.23), (3.42) and (2.38),

$$\sigma_{xx} \left(= \frac{\partial^2 \phi}{\partial y^2} = f_{(x)}\, g''_{(y)} \right) \propto (Y + K_1 + K_2 y) \tag{3.44}$$

where K_1 and K_2 are constants determined from (2.38) when Y is known. Thus if $g''_{(y)}$ is taken to be

$$g''_{(y)} = Y + K_1 + K_2 y \tag{3.45}$$

successive integration yields expressions for $g'_{(y)}$ and $g_{(y)}$. The various boundary conditions on the stresses are (using (2.23))

$$\left.\begin{aligned} &\sigma_{xx} = fg'' = 0\,; \quad \sigma_{xy} = -f'g' = 0 \text{ on } x = \pm\, a, \\ \text{and} \quad &\sigma_{yy} = f''g = 0\,; \quad \sigma_{xy} = -f'g' = 0 \text{ on } y = \pm\, b \end{aligned}\right\} \tag{3.46}$$

Thus the expressions for g and g' must both equal zero when $y = \pm\, b$.

Having obtained an expression for g it then remains to determine the function f.

Because the plate is unrestrained, the complementary strain energy, eqn. (1.47), may be used for this problem. This, in the present notation, becomes

$$\begin{aligned} 2E\bar{U}_0' = \int_{-a}^{a} \{ & A_1 f^2 + A_2 (f'')^2 - 2\nu A_3 f f'' + 2(1 + \nu) A_4 (f')^2 \\ & + 2E\alpha[(A_5 + A_6 X) f + (A_7 + A_8 X) f''] \} \mathrm{d}x \end{aligned} \tag{3.47}$$

where $A_1 = \int_{-b}^{b} (g'')^2 \mathrm{d}y$ $\qquad A_5 = T_0[g']_{-b}^{b}$

$$A_2 = \int_{-b}^{b} g^2\, \mathrm{d}y \qquad A_6 = \int_{-b}^{b} Y g'' \mathrm{d}y$$

$$A_3 = \int_{-b}^{b} gg'' \mathrm{d}y = [gg']_{-b}^{b} - A_4 \qquad A_7 = T_0 \int_{-b}^{b} g\, \mathrm{d}y$$

$$A_4 = \int_{-b}^{b} (g')^2 \mathrm{d}y \qquad A_8 = \int_{-b}^{b} Y g\, \mathrm{d}y$$

The expression for $\overline{U}_0'$ is converted into an ordinary differential equation by using Euler's equation; hence, by making use of the boundary conditions (3.46) to simplify some of the A_m terms above, the resulting equation is

$$A_2 f^{iv} - 2A_4 f'' + A_1 f = E\alpha(A_6 X + A_8 X'') \tag{3.48}$$

For the temperature distribution of eqn. (3.35) $X = -\Delta T$, which is constant, so that $X'' = 0$ in (3.48); and

$$g'' = \left|\frac{y}{b}\right|^n - \frac{1}{n+1} \tag{3.49}$$

Substitution of these expressions into eqn. (3.48) gives the particular result,[143]

$$B_1 f^{iv} - B_2 f'' + f = E\alpha\Delta T \tag{3.50}$$

where $B_1 = 2b^4(2n + 1)(2n + 11)/15(n + 3)(n + 5)(2n + 5)$,
and $B_2 = 4b^2(2n + 1)/3(n + 3)(2n + 3)$

The general solution of (3.50) using the appropriate boundary conditions of (3.46) is,

$$f = E\alpha\Delta T[1 + C_1 \sinh k_1x \sin k_2x + C_2 \cosh k_1x \cos k_2x] \tag{3.51}$$

where,

$$DC_1 = k_1 \sinh k_1a \cos k_2a - k_2 \cosh k_1a \sin k_2a$$

$$DC_2 = -k_1 \cosh k_1a \sin k_2a - k_2 \sinh k_1a \cos k_2a$$

$$D = k_1 \sin k_2a \cos k_2a + k_2 \sinh k_1a \cosh k_1a$$

$$k_1^2 = [B_2 + 2\sqrt{B_1}]/4B_1$$

$$k_2^2 = [-B_2 + 2\sqrt{B_1}]/4B_1$$

The result for g is

$$g = b^2 \left[n - (n+2) \left| \frac{y}{b} \right|^2 + 2 \left| \frac{y}{b} \right|^{n+2} \right] / 2(n+1)(n+2) \qquad (3.52)$$

Hence, knowing $\phi = fg$ for any plate geometry and value of n the expressions for the thermal stress distributions can be found from (2.23). A comparison of this method with Method 1 is shown in Fig. 3.3 for the case of $n = 1$, $a/b = 3/2$, which was investigated

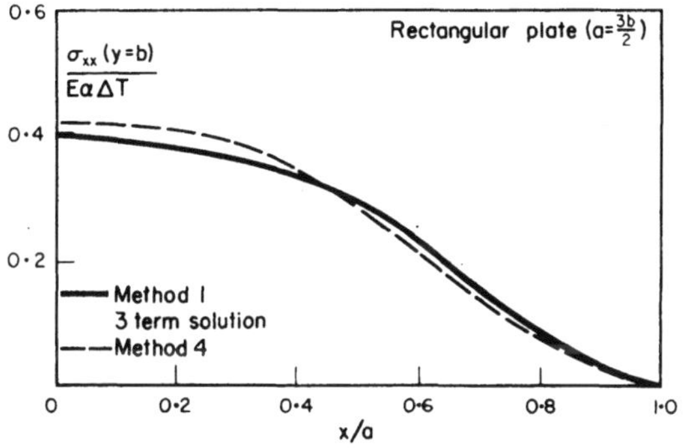

FIG. 3.3. Thermal stress along edge of rectangular plate for a " linear " (tent-like) temperature distribution—Taken from reference 143. (*J. Aero. Soc. of India*).

initially, both theoretically and experimentally in Ref. 62. The agreement between the two methods is satisfactory.

It may be concluded from these studies that for one-dimensional temperature distributions Method 1 is adequate and probably simpler to apply. For two-dimensional temperature distributions the choice of methods is not obvious, although in reference 144 it is claimed that the collocation procedure[115] pursued with the aid of a high speed computer is more adaptable to a wider range of problems.

Further studies of this problem are contained in references 94 and 118, in which allowance is made for the presence of supporting structure attached to the plate.

CHAPTER 4

Bending Thermal Stresses in Thin Plates

4.1. Theory for Thin Isotropic Plates with Small Deflexions

The basic assumptions of small deflexion thin plate theory are:

(1) Points of the plate which before bending lie on the normal to the median plane of the plate remain on the normal after bending, i.e. $e_{xz} = e_{yz} = 0$.

(2) The elongation of this normal is negligible, i.e. $e_{zz} \simeq 0$.

(3) Normal stresses through the plate thickness are negligible, i.e. $\sigma_{zz} \simeq 0$.

(4) The plate is homogeneous and isotropic with temperature-independent material properties. (Modifications to the following analyses to allow for temperature dependent properties are given in reference 127 and discussed in Chapter 8.)

From the three-dimensional formulations of Chapter 1 various simplifications can be made using assumptions (1) and (2), viz.

if
$$e_{xz} = 0; \quad e_{yz} = 0; \quad e_{zz} = 0 \tag{4.1}$$

then from (1.9)

$$\frac{\partial u}{\partial z} = -\frac{\partial w}{\partial x}; \quad \frac{\partial v}{\partial z} = -\frac{\partial w}{\partial y}; \quad w = w_{o(x,y)} \tag{4.2}$$

Hence,

$$\left.\begin{aligned} u &= u_{o(x,y)} - z\frac{\partial w}{\partial x} \\ v &= v_{o(x,y)} - z\frac{\partial w}{\partial y} \\ w &= w_{o(x,y)} \end{aligned}\right\} \tag{4.3}$$

and the corresponding expressions for the strains in the plane of the plate, in terms of displacements and stresses are,

$$e_{xx} = \frac{\partial u_o}{\partial x} - z\frac{\partial^2 w}{\partial x^2} = \frac{1}{E}\left[\sigma_{xx} - \nu\sigma_{yy}\right] + \alpha T \tag{4.4}$$

$$e_{yy} = \frac{\partial v_o}{\partial y} - z\frac{\partial^2 w}{\partial y^2} = \frac{1}{E}\left[\sigma_{yy} - \nu\sigma_{xx}\right] + \alpha T \tag{4.5}$$

$$e_{xy} = \frac{\partial u_o}{\partial y} + \frac{\partial v_o}{\partial x} - 2z\frac{\partial^2 w}{\partial x \partial y} = \frac{2(1+\nu)}{E}\sigma_{xy} \tag{4.6}$$

where suffix o indicates values of the displacements at the median plane of the plate.

The corresponding non-zero stress components for this plane stress problem are,

$$\left.\begin{aligned} \sigma_{xx} &= \frac{E}{1-\nu^2}\left[e_{xx} + \nu e_{yy} - (1+\nu)\,\alpha T\right] \\ \sigma_{yy} &= \frac{E}{1-\nu^2}\left[e_{yy} + \nu e_{xx} - (1+\nu)\,\alpha T\right] \\ \sigma_{xy} &= \frac{E}{2(1+\nu)}e_{xy} \end{aligned}\right\} \tag{4.7}$$

and integrating these over the plate thickness to obtain the appropriate force and moment resultants[164] yields, ($N = \int \sigma dz$, $M = \int \sigma z dz$)

$$\left.\begin{aligned} N_x &= \frac{Ed}{1-\nu^2}\left(\frac{\partial u_o}{\partial x} + \nu\frac{\partial v_o}{\partial y}\right) - \frac{N_T}{1-\nu} \\ N_y &= \frac{Ed}{1-\nu^2}\left(\frac{\partial v_o}{\partial y} + \nu\frac{\partial u_o}{\partial x}\right) - \frac{N_T}{1-\nu} \\ N_{xy} &= \frac{Ed}{2(1+\nu)}\left(\frac{\partial u_o}{\partial y} + \frac{\partial v_o}{\partial x}\right) \end{aligned}\right\} \tag{4.8}$$

$$\left.\begin{aligned} M_x &= -D\left(\frac{\partial^2 w}{\partial x^2} + \nu\frac{\partial^2 w}{\partial y^2}\right) - \frac{M_T}{1-\nu} \\ M_y &= -D\left(\frac{\partial^2 w}{\partial y^2} + \nu\frac{\partial^2 w}{\partial x^2}\right) - \frac{M_T}{1-\nu} \\ M_{xy} &= D(1-\nu)\frac{\partial^2 w}{\partial x \partial y} \end{aligned}\right\} \tag{4.9}$$

In these expressions D is the bending rigidity of the plate, whose local thickness is d, and where

$$D = Ed^3/12(1 - \nu^2) \tag{4.10}$$

and the quantities N_T and M_T are given by

$$N_T = E\alpha \int_{-d/2}^{d/2} T\,\mathrm{d}z, \quad M_T = E\alpha \int_{-d/2}^{d/2} Tz\,\mathrm{d}z \tag{4.11}$$

The equations of equilibrium in the plane of the plate, neglecting body forces, are given by

$$\frac{\partial N_x}{\partial x} + \frac{\partial N_{xy}}{\partial y} = 0\ ; \quad \frac{\partial N_{xy}}{\partial x} + \frac{\partial N_y}{\partial y} = 0 \tag{4.12}$$

and the existence of a stress function, ϕ_N is therefore implied such that

$$N_x = +\frac{\partial^2 \phi_N}{\partial y^2}, \quad N_{xy} = -\frac{\partial^2 \phi_N}{\partial x \partial y}, \quad N_y = +\frac{\partial^2 \phi_N}{\partial x^2} \tag{4.13}$$

By using eqn. (4.8), the equation of compatibility (2.18) written in terms of median plane strains $(e_{xx})_o$ etc., becomes[15]

$$\frac{\partial^2}{\partial y^2}\left[\frac{N_x - \nu N_y + N_T}{d}\right] - 2(1+\nu)\frac{\partial^2}{\partial x \partial y}\left[\frac{N_{xy}}{d}\right] + \frac{\partial^2}{\partial x^2}\left[\frac{N_y - \nu N_x + N_T}{d}\right] = 0 \tag{4.14}$$

This may be simplified for plates of uniform thickness, and by using (4.13), to give

$$\nabla^4 \phi_N = -\nabla^2 N_T\ ; \quad \nabla^2 = \frac{\partial^2}{\partial x^2} + \frac{\partial^2}{\partial y^2} \tag{4.15}$$

This equation resembles (2.24) derived in Chapter 2 and is of no further concern in the present bending problem. Its solution leads only to membrane stresses which do not influence the bending behaviour of the plate for small deflexions.

The equations of equilibrium for forces in the z direction and for moments about the x and y axes are,[164] neglecting body forces,

$$\left.\begin{aligned}\frac{\partial Q_x}{\partial x}+\frac{\partial Q_y}{\partial y}+p&=0\\ \frac{\partial M_{xy}}{\partial x}-\frac{\partial M_y}{\partial y}+Q_y&=0\\ \frac{\partial M_{xy}}{\partial y}-\frac{\partial M_x}{\partial x}+Q_x&=0\end{aligned}\right\}\tag{4.16}$$

where $p = p_{(x,y)}$ is a transverse pressure distribution normal to the median plane of the plate and Q_x, Q_y are the shear force resultants given by

$$Q_x = \int_{-d/2}^{d/2} \sigma_{xz} dz \neq 0 \ ; \ Q_y = \int_{-d/2}^{d/2} \sigma_{yz} dz \neq 0 \tag{4.17}$$

The equation of equilibrium for forces in the z direction has the following form, when the terms in Q_x and Q_y are eliminated by substitution from the remaining equations of (4.16),

$$\frac{\partial^2 M_x}{\partial x^2} - 2\frac{\partial^2 M_{xy}}{\partial x \partial y} + \frac{\partial^2 M_y}{\partial y^2} = -p \tag{4.18}$$

By using (4.9) this equation becomes,

$$\frac{\partial^2}{\partial x^2}\left[D\left(\frac{\partial^2 w}{\partial x^2}+\nu\frac{\partial^2 w}{\partial y^2}\right)\right] + 2(1-\nu)\frac{\partial^2}{\partial x \partial y}\left[D\frac{\partial^2 w}{\partial x \partial y}\right]$$
$$+\frac{\partial^2}{\partial y^2}\left[D\left(\frac{\partial^2 w}{\partial y^2}+\nu\frac{\partial^2 w}{\partial x^2}\right)\right] = p - \frac{1}{1-\nu}\nabla^2 M_T \tag{4.19}$$

which may be further simplified, if the plate has uniform thickness, to give,

$$D\nabla^4 w = p - \frac{1}{1-\nu}\nabla^2 M_T \tag{4.20}$$

It is clear, therefore, that eqns. (4.19) and (4.20) are equivalent to their isothermal forms with a normal loading of $p - \nabla^2 M_T/(1-\nu)$. Therefore all the known techniques for isothermal problems may be used also for the corresponding thermoelastic problems.

In the derivation above it was implicitly assumed that $T = T_{(x,y,z)}$, but, when the temperature varies arbitrarily in the thickness direction of the plate (z) as well as over the planform (x, y) then ideally the general equations of three-dimensional thermoelasticity should be used. For this reason the only admissible forms of temperature distribution which do not invalidate the plane stress hypothesis are, (see Sections 2.6 and 2.2),

$$T = T_{(z)} \tag{4.21}$$

i.e. variations through the thickness only, and

$$T = T_{o(x,y)} - \frac{z}{d}\Delta T_{(x,y)} \tag{4.22}$$

i.e. variations over the planform (x, y) which are linear in z.

The distribution of (4.21) gives constant values of N_T and M_T resulting in obvious simplification in the solution of eqns. (4.15) and (4.20). The distribution of (4.22) gives N_T dependent only on T_o and M_T dependent only on ΔT. Since the term T_o only affects the membrane thermal stresses, which were considered in Chapter 3 they will be omitted from further discussion. The distributions $T = T_{(z)}$ and $T = -z\Delta T_{(x,y)}/d$ will receive separate consideration in Sections 4.4 and 4.5.

4.2. Theory for Thin Isotropic Plates with Large Deflexions

When the plate deflexions are not small compared with the plate thickness there is a coupling between the membrane and bending actions of the plate which evinces itself in two important respects. The strains in the plane of the plate are found to be dependent on non-linear terms in the deflexion w; and the equation of equilibrium for forces in the z direction contains significant terms from the components of the force resultants, N, in the distorted median plane.

Therefore, the large-deflexion strain–displacement relations are, (cf. with (4.4)–(4.6))

$$\left.\begin{aligned}
e_{xx} &= \frac{\partial u_o}{\partial x} - z\frac{\partial^2 w}{\partial x^2} + \frac{1}{2}\left(\frac{\partial w}{\partial x}\right)^2 \\
e_{yy} &= \frac{\partial v_o}{\partial y} - z\frac{\partial^2 w}{\partial y^2} + \frac{1}{2}\left(\frac{\partial w}{\partial y}\right)^2 \\
e_{xy} &= \frac{\partial u_o}{\partial y} + \frac{\partial v_o}{\partial x} - 2z\frac{\partial^2 w}{\partial x \partial y} + \frac{\partial w}{\partial x}\frac{\partial w}{\partial y}
\end{aligned}\right\} \tag{4.23}$$

and the only changed equilibrium equation reads (cf. with (4.16))

$$\frac{\partial Q_x}{\partial x} + \frac{\partial Q_y}{\partial y} + p + N_x \frac{\partial^2 w}{\partial x^2} + N_y \frac{\partial^2 w}{\partial y^2} + 2N_{xy} \frac{\partial^2 w}{\partial x \partial y} = 0 \qquad (4.24)$$

Following the procedure of the previous section the governing equation which is obtained for w, corresponding to (4.20), is,

$$D\nabla^4 w = p - \frac{1}{1-\nu}\nabla^2 M_T + \left[\frac{\partial^2 \phi_N}{\partial y^2}\frac{\partial^2 w}{\partial x^2} + \frac{\partial^2 \phi_N}{\partial x^2}\frac{\partial^2 w}{\partial y^2} - 2\frac{\partial^2 \phi_N}{\partial x \partial y}\frac{\partial^2 w}{\partial x \partial y}\right] \qquad (4.25)$$

where N_x, N_y and N_{xy} have been written in terms of the stress function ϕ_N, i.e. using (4.13). The compatibility equation is derived as before, but using (4.23), to yield, (cf. with (4.15))

$$\nabla^4 \phi_N + \nabla^2 N_T = Ed\left[\left(\frac{\partial^2 w}{\partial x \partial y}\right)^2 - \frac{\partial^2 w}{\partial x^2}\cdot\frac{\partial^2 w}{\partial y^2}\right] \qquad (4.26)$$

Therefore the large deflexion problem is reduced to the solution of two simultaneous, non-linear, differential equations (4.25) and (4.26) containing two unknown functions $w_{(x,y)}$ and $\phi_{N(x,y)}$. The solution of these equations for general problems is not, unfortunately, available, but approximate solutions exist and are fully discussed in reference 164.

The results above may be modified in certain cases, as follows.

(1) If the plate is very thin and the deflexions are many times larger than the thickness, the plate effectively becomes a membrane with zero bending rigidity, i.e. $D = 0$ in eqn. (4.25).

(2) If the median plane forces are large but the bending deflexions are very small, then eqns. (4.25) and (4.26) may be uncoupled by putting the right side of eqn. (4.26) equal to zero. This equation is then solved for ϕ_N which is substituted into eqn. (4.25). This condition corresponds to the assumption usually made in plate buckling analyses; that the membrane forces causing buckling do not change during the buckling process (see Chapter 7).

(3) If the bending deflexions are small, but the plate has a small initial curvature, eqns. (4.25) and (4.26) are modified as follows,

$$D\nabla^4 w_1 = p - \frac{1}{1-\nu}\nabla^2 M_T + \left[\frac{\partial^2\phi_N}{\partial y^2}\frac{\partial^2 w}{\partial x^2} + \frac{\partial^2\phi_N}{\partial x^2}\frac{\partial^2 w}{\partial y^2} - 2\frac{\partial^2\phi_N}{\partial x\partial y}\frac{\partial^2 w}{\partial x\partial y}\right] \tag{4.27}$$

$$\nabla^4\phi_N + \nabla^2 N_T = Ed\left[\left(\frac{\partial^2 w}{\partial x\partial y}\right)^2 - \frac{\partial^2 w}{\partial x^2}\frac{\partial^2 w}{\partial y^2}\right] - Ed\left[\left(\frac{\partial^2 w_o}{\partial x\partial y}\right)^2 - \frac{\partial^2 w_o}{\partial x^2}\frac{\partial^2 w_o}{\partial y^2}\right] \tag{4.28}$$

where w is the total deflexion of the middle surface

w_o is the initial deflexion of the middle surface in the unloaded and unheated state

w_1 is the additional deflexion of the middle surface due to loading and/or heating,

i.e. $w = w_1 + w_o$ where $w_o \ll d$.

As in Case 2, when the bending deflexions are very small, the two equations above may be uncoupled by setting the right-hand side of (4.28) equal to zero.

4.3. Boundary Conditions

The boundary conditions to be satisfied in the solution of plate bending problems are the same for both small and large deflexion analyses viz. there are two conditions, on ϕ_N and its derivative, to be satisfied for the membrane behaviour of the plate; and there are two conditions on w and its derivatives concerning the bending behaviour of the plate.

For the membrane problem, the appropriate boundary conditions have been discussed and used already in Chapters 2 and 3. Suffice to say, that for traction-free surfaces $\phi_N = \partial\phi_N/\partial n = 0$, whereas for specified edge tractions the appropriate relationships are given by (4.13).

For the bending problem there are four distinct types of edge condition.

I. At a *built-in* edge,

$$w = \frac{\partial w}{\partial n} = 0 \tag{4.29}$$

II. At a *simply-supported* edge,

$$w = M_n = 0 \tag{4.30}$$

where M_n may be written more fully as

$$-M_n = \nu D \nabla^2 w + (1-\nu) D \left(l^2 \frac{\partial^2 w}{\partial x^2} + 2lm \frac{\partial^2 w}{\partial x \partial y} + m^2 \frac{\partial^2 w}{\partial y^2} \right) + \frac{M_T}{1-\nu} \tag{4.31}$$

III. At a *free* edge, of a continuous curvilinear boundary, the requirements to be satisfied are those of zero bending and twisting moments and zero vertical shear force. For a boundary having corners the *reactions* at those corners must also be zero.[164] Expressed symbolically these conditions reduce to become,

$$M_n = 0, \quad \frac{\partial M_n}{\partial n} - 2 \frac{\partial M_{ns}}{\partial s} = 0, \quad R_c = (M_{ns})_1 - (M_{ns})_2 = 0 \tag{4.32}$$

where subscripts 1 and 2 refer to values of M_{ns} on the sides forming the corner. The twisting moment M_{ns} may be written more fully as

$$M_{ns} = (1-\nu) D \left[\frac{\partial^2 w}{\partial x \partial y} (l^2 - m^2) + lm \left(\frac{\partial^2 w}{\partial y^2} - \frac{\partial^2 w}{\partial x^2} \right) \right] \tag{4.33}$$

and the following relations should also be noted for use in (4.29) and (4.32), where l and m are direction cosines,

$$\frac{\partial}{\partial n} = l \frac{\partial}{\partial x} + m \frac{\partial}{\partial y}\, ; \quad \frac{\partial}{\partial s} = l \frac{\partial}{\partial y} - m \frac{\partial}{\partial x} \tag{4.34}$$

IV. For *elastic restraint* relations are set up between corresponding pairs of force and displacement vectors using information already given.

4.4. Temperature Distributions, $T = T_{(z)}$

4.4.1. Free Plate of Arbitrary Planform; $T = T_{(z)}$[15]

When the temperature varies in the z direction only, and is independent of x and y, both N_T and M_T are constants and a simple set of results is obtained for the small deflexion problem.

With N_T constant the solution to eqn. (4.15) which also satisfies the condition of traction-free edges is $\phi_N = 0$. Therefore $N_x = N_y = N_{xy} = 0$ and from eqn. (4.8) the median plane displacements, u_o and v_o, are obtained, viz.

$$\left.\begin{aligned} u_{o(x,y)} &= N_T \,.\, x/Ed + [a + by] \\ v_{o(x,y)} &= N_T \,.\, y/Ed + [c - bx] \end{aligned}\right\} \tag{4.35}$$

With M_T constant and $p = 0$ the solution to eqn. (4.20) which also satisfies the free edge conditions is

$$w_{(x,y)} = \frac{-6M_T}{Ed^3}(x^2 + y^2) + [e + fx + gy] \tag{4.36}$$

The terms in the square brackets correspond to rigid body motions. The resulting expressions for the direct stresses are identical to those presented in Section 2, eqn. (2.41), viz.

$$\sigma_{xx(z)} = \sigma_{yy(z)} = \frac{1}{1-\nu}\left[\frac{N_T}{d} + \frac{12M_T z}{d^3} - E\alpha T\right] \tag{4.37}$$

If the plate is restrained against expansion in its plane the term N_T is omitted and if the plate is restrained against edge rotation the term in M_T is omitted, leaving, for the fully restrained plate, stresses in the plane of the plate given by

$$\sigma_{xx(z)} = \sigma_{yy(z)} = -E\alpha T/(1-\nu) \tag{4.38}$$

For the present chapter attention is devoted to the term in M_T which enters directly into the plate bending equation (4.20). Thus, all plate bending solutions depend on the quantity M_T defined in (4.11). Since the effect of M_T is equivalent to a simple temperature gradient, ΔT, across the plate thickness, this latter quantity will be used in the remainder of Section 4.4. To replace ΔT by M_T in the analyses of this section, the relevant relationship is $\Delta T = 12M_T/E\alpha d^2$ where both ΔT and M_T are constant over the plate. ΔT is defined so that the plate upper surface ($z = -d/2$) is hotter than the lower ($z = d/2$).

4.4.2. Circular Plates; $T_{(z)} = \Delta T(z/d)$

An unrestrained circular plate experiences a radius of curvature $R = d/\alpha\Delta T$ and, since the entire edge of the spherical cap so formed remains in the same plane, no stresses are induced with a *simply supported* edge condition.

If the edges are clamped however, edge moments are set up so as to cancel the curvature caused by the thermal deformation of the unrestrained plate. These moments are related, through R, to ΔT by the equation,

$$M_n = a\,\Delta TD\,(1 + \nu)/d \tag{4.39}$$

which applies to all clamped plates irrespective of their planform. Therefore the maximum stresses are

$$(\sigma_{rr})_{\max} = (\sigma_{\theta\theta})_{\max} = \pm\, Ea\Delta T/2(1 - \nu) \tag{4.40}$$

and the transverse deformations in this case are zero. It can be deduced that since the expressions for the maximum stresses are independent of the plate thickness, but ΔT would be expected to increase with plate thickness, then greater thermal stresses can be expected in thick plates than in thin ones.

Solutions for hollow circular plates with various boundary conditions on the concentric circumferential edges are given in a series of papers by Newman and Forray(126).

4.4.3. Rectangular Plates; $T_{(z)} = -\Delta T(z/d)$

For rectangular plates restrained against rotation at all four edges an identical result is obtained to the clamped circular plate, viz. the transverse deflexions are zero and

$$(\sigma_{xx})_{\max} = (\sigma_{yy})_{\max} = \pm\, Ea\Delta T/2(1 - \nu) \tag{4.41}$$

To obtain the thermal stress distributions for the simply supported plate it is only necessary to superpose on the uniformly distributed moments M_n of (4.39), the moments produced in the plate with simply-supported edges by uniformly distributed moments $-M_n$ along the plate edges.

For this latter problem, the application of $-M_n$ satisfies the condition of zero edge moment and the only further restriction is that the deflexion surface should have zero values at the edges. In fact the plate deflexions must satisfy the equation,

$$\frac{\partial^2 w}{\partial x^2} + \frac{\partial^2 w}{\partial y^2} = a\Delta T(1 + \nu)/d \tag{4.42}$$

The solution to (4.42) is found as an infinite trigonometric series, as are the expressions for the bending moment distributions. The final result is obtained by summing these bending moment distributions with the expression in eqn. (4.39). The results as obtained by Maulbetsch[110] and presented also in reference 164 may be written as follows,

$$w = \frac{-\alpha\Delta T(1+\nu)4a^2}{\pi^3 d}\sum_{m=1,3,5,\ldots}^{\infty} \sin\frac{m\pi x}{a}\left(1-\frac{\cosh\frac{m\pi y}{a}}{\cosh \alpha_m}\right)\Big/ m^3 \tag{4.43}$$

$$M_x = \frac{4D\alpha\Delta T(1-\nu^2)}{\pi d}\sum_{m=1,3,5,\ldots}^{\infty} \sin\frac{m\pi x}{a}\,.\,\cosh\frac{m\pi y}{a}\Big/ m\cosh\alpha_m \tag{4.44}$$

$$M_y = \frac{D\alpha\Delta T(1-\nu^2)}{d} - \frac{4D\alpha\Delta T(1-\nu^2)}{\pi d}\times \sum_{m=1,3,5,\ldots}^{\infty} \sin\frac{m\pi x}{a}\cosh\frac{m\pi y}{a}\Big/ m\cosh\alpha_m \tag{4.45}$$

$$M_{xy} = \frac{4D\alpha\Delta T(1-\nu^2)}{\pi d}\sum_{m=1,3,5,\ldots}^{\infty} \cos\frac{m\pi x}{a}\sinh\frac{m\pi y}{a}\Big/ m\cosh\alpha_m \tag{4.46}$$

where $\alpha_m = m\pi b/2a$, and a, b are the plate dimensions in the coordinate directions x, y which are based at the centre of a side of length b. It should be noted that if $D = Ed^3/12(1-\nu^2)$ is substituted into the bending moment expressions they are then independent of Poisson's ratio, and only depend on the product $E\alpha\Delta T$ and on the plate aspect ratio (a/b). Figure 4.1 shows the variation of the thermal stresses at the centre of a rectangular plate (when $x = a/2$, $y = 0$) with aspect ratio.[142] The maximum stress at the centre $(\pm E\alpha\Delta T/2)$ is only experienced when $a/b \rightarrow \infty$, and the minimum stress at the centre $(\pm E\alpha\Delta T/4)$ occurs on the square plate $(a/b = 1)$. However, it is found that the absolute maximum values of thermal stress occur at the boundary of the plate and are independent of the aspect ratio, i.e.

$$\left(\sigma_{xx}\right)_{y = \pm b/2} = \left(\sigma_{yy}\right)_{x = 0,\, a} = \pm E\alpha\Delta T/2 \tag{4.47}$$

This result could have been found more directly in the following way. Due to the application of $-M_n$ along a simply supported edge a moment $M_t = -\nu M_n$ must act in the direction of the edge. This moment acts in addition to the moments M_n on the originally clamped plate so that the total moment acting in the direction of a

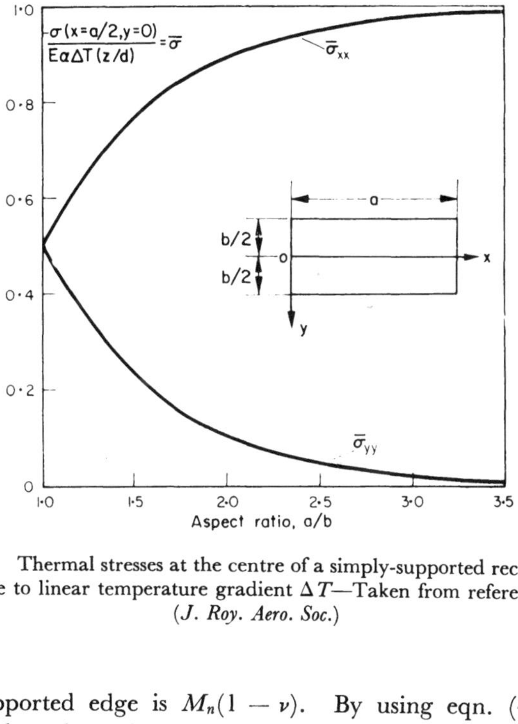

FIG. 4.1. Thermal stresses at the centre of a simply-supported rectangular plate due to linear temperature gradient ΔT—Taken from reference 142. (*J. Roy. Aero. Soc.*)

simply-supported edge is $M_n(1 - \nu)$. By using eqn. (4.39) this results in the values of stress quoted in (4.47).

The analyses of references 110 and 164 indicate that the concentrated reactions, R_c, and twisting moments, M_{xy}, tend to infinity at the sharp corners of the plate. In ductile materials this would give rise to local yielding and cause a redistribution of stress near the corners. In practice these values can be made finite by the redistribution of the concentrated reactions into, say, a bolt group or by rounding the corners. A method of analysis for this modification is given in references 110, 12.

This latter paper deals specifically with sandwich panels for which it is assumed that the flexural rigidity D can be replaced by

$$D_s = Ed_F(d_c + d_F)^2/2(1 - \nu^2) \tag{4.48}$$

where d_c is the thickness of the sandwich core and d_F is the thickness of each facing, i.e. overall thickness is $d_c + 2d_F$. Various combinations of edge support conditions are considered and it is shown that with the modified value of D_s similar results are obtained as for homogeneous plates—when either clamped (built-in) or simply

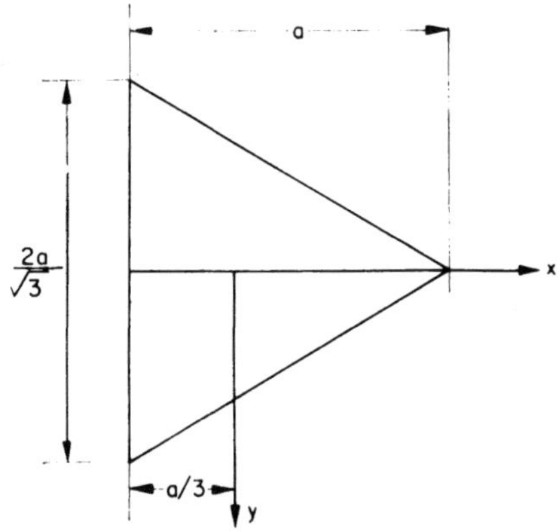

FIG. 4.2. Equilateral triangular plate.

supported edges are assumed. For other cases the shear deformation of the core is significant and must be included whereas the corresponding term for homogeneous plates is not usually considered.[164]

4.4.4. Polygonal Plates; $T_{(z)} = -\Delta T(z/d)$

It should be noted that polygonal planforms other than rectangular can also be analysed by the above procedure. Thus for a clamped plate the result of (4.39) again applies, and for a simply supported plate a cancelling moment $-M_n$ is assumed and its effect determined as a problem of isothermal elasticity.

Results for the simply-supported, equilateral, triangular plate of Fig. 4.2, have a particularly simple form, viz.[110,164]

$$\left.\begin{aligned} M_x &= E\alpha\Delta Td^2\left(1+\frac{3x}{a}\right)\Big/24 \\ M_y &= E\alpha\Delta Td^2\left(1-\frac{3x}{a}\right)\Big/24 \\ M_{xy} &= E\alpha\Delta Td^2 y/8a \\ R_c &= -\,E\alpha\Delta Td^2/4\sqrt{3} \end{aligned}\right\} \tag{4.49}$$

4.4.5. Sundry Solutions; $T = T_{(z)}$

Other solutions to the present type of problem are available in the literature and a selection will now be given. The differences between these solutions are in either the edge restraints considered or in the assumed heating environment. Gatewood[48] has presented several of these solutions in greater detail. For information of the heat transfer terminology used, reference should be made to the Appendix.

I. Plate with prescribed linear transient temperature on one surface ($z = d/2$), i.e.

$$\left.\begin{aligned} T_s &= T_E\, t/t_1 \qquad & 0 \leqslant t \leqslant t_1 \\ T_s &= T_E & t \geqslant t_1 \end{aligned}\right\} \tag{4.50}$$

An important parameter on which the thermal stresses depend is the dimensionless quantity, $W = \kappa t_1/d^2$. As this quantity becomes smaller, i.e. for a thicker plate, or a shorter time rise, or a lower diffusivity, the stresses become larger.

II. Plate with constant convective heat source on one surface ($z = d/2$)

In the notation of the Appendix this condition is expressed by

$$h(T_{AW} - T_s) = k\,\frac{\partial T}{\partial z} \tag{4.51}$$

With the heat transfer coefficient, h, and adiabatic wall temperature, T_{AW}, both constant this case corresponds to the problem of convective heating into a *thermally thick* plate and is considered in detail by Przemieniecki[139]. There are two parameters on which the values of thermal stress depend most strongly, viz.

$$W = \kappa t/d^2 \text{ and } B = hd/k \text{ (the Biot number).}$$

III. Plates with variable heat source on one surface.

For convective heating, eqn. (4.51) is used but h and/or T_{AW} may vary with time. Heisler[59] assumes h to be constant and takes the following variation of T_{AW}, i.e.

$$\left.\begin{aligned} T_{AW} &= T_E\, t/t_o \qquad 0 \leqslant t \leqslant t_o \\ T_{AW} &= T_E \qquad\qquad t \geqslant t_o \end{aligned}\right\} \tag{4.52}$$

A variable heat input of the form,

$$q = gte^{-t/\beta} \tag{4.53}$$

with g and β constant, is considered by Weiner and Mechanic[172] to be applied to one face ($z = +d/2$) of a free plate. Equation (4.53) can represent the form of heat input experienced due to atomic explosion. The stress is expressed non-dimensionally by

$$\psi = \sigma_{xx}k\kappa(1 - \nu)E\alpha g d^3 \tag{4.54}$$

and is found to be strongly dependent on the two parameters $W = \kappa t/d^2$ and $S = \kappa\beta/d^2$. It is found that the maximum stress in the plate can be written approximately as

$$\psi_{max} = \cdot 0153S \tag{4.55}$$

4.5. Temperature Distributions, $T = -z\Delta T_{(x,y)}/d$

The derivations of Section 4.1 have been shown to be applicable to temperature distributions of the form of (4.22), since the plane stress hypothesis for thin plates is not, thereby, violated.

Omitting the lateral pressure term, p, from eqn. (4.20) and substituting $T = -z\Delta T_{(x,y)}/d$, results in,

$$\nabla^4 w = +\frac{(1+\nu)}{d}\nabla^2(\alpha\Delta T) \tag{4.56}$$

This equation has the same form as (2.24) and (4.15) and it would suggest that there should be a resemblance between the analyses for membrane thermal stresses (Chapter 3) and those for the present bending problem. Such a similarity is observed later. Solutions of this equation for rectangular plates are given in reference 22.

4.5.1. Axisymmetric Bending of Hollow Circular Plates

If ΔT is a function of the radial coordinate only in a hollow circular plate, the solution to (4.56) is obtained by successive integration as

$$w = + \frac{(1+\nu)}{d}\left[\int_a^r \frac{F_{(r)}}{r}\,\mathrm{d}r + C_1 + C_2 \log r + C_3 r^2 + C_4 r^2 \log r\right] \tag{4.57}$$

where $F_{(r)} = \int_a^r \alpha\Delta T r \mathrm{d}r$ and $C_1 \ldots C_4$ are arbitrary constants to be determined by the boundary conditions on the inner and outer faces of the annular plate, i.e. on $r = a$, $r = b$. The boundary conditions of Section 4.3 need to be expressed in polar coordinates for easier determination of the constants C_1 to C_4.

The bending moments per unit length, M_r and M_θ, are given by eqn. (4.9).

$$M_r = -D\left(\frac{\mathrm{d}^2 w}{\mathrm{d}r^2} + \frac{\nu}{r}\frac{\mathrm{d}w}{\mathrm{d}r}\right) + \frac{Ed^2\alpha\Delta T}{12(1-\nu)} \tag{4.58}$$

$$M_\theta = -D\left(\frac{1}{r}\frac{\mathrm{d}w}{\mathrm{d}r} + \nu\frac{\mathrm{d}^2 w}{\mathrm{d}r^2}\right) + \frac{Ed^2\alpha\Delta T}{12(1-\nu)} \tag{4.59}$$

where M_r and M_θ are + ve when the upper fibres are in compression; and the temperature of the upper surface is greater than that of the lower surface.

For a simply supported or free solid plate ($a = 0$) the appropriate boundary conditions are $w = M_r = 0$ at $r = b$; in order to avoid singularities at $r = 0$ the constants C_2 and C_4 must be zero. The results which may easily be verified are

$$w = \frac{1+\nu}{d}\left[\int_0^r \frac{F_{(r)}}{r}\mathrm{d}r - \int_0^b \frac{F_{(r)}}{r}\mathrm{d}r - \frac{(1-\nu)}{2(1+\nu)}F_{(b)}\left(1 - \frac{r^2}{b^2}\right)\right] \tag{4.60}$$

$$M_r = \frac{Ed^2}{12}\left[\frac{1}{r^2}F_{(r)} - \frac{1}{b^2}F_{(b)}\right] \tag{4.61}$$

$$M_\theta = \frac{Ed^2}{12}\left[\alpha\Delta T - \frac{1}{r^2}F_{(r)} - \frac{1}{b^2}F_{(b)}\right] \tag{4.62}$$

It follows from this analysis that there is zero radial extension to the plate but the rotation of the circumferential edge may be written as

$$\theta_b = 2F_{(b)}/bd \tag{4.63}$$

The extreme fibre stresses are given by $\sigma = 6M/d^2$ and are independent of the thickness parameter, d, but since ΔT may be larger for larger d, the thermal stresses may therefore *increase* with plate thickness.

Results for various boundary conditions are available in the literature. Simply supported and clamped solid plates are discussed generally in reference 56; and in reference 42 formulae are given for specific polynomial forms of radial temperature distribution.

Thus if $\Delta T = a_K r^K$ the results for a solid clamped plate are[42]

$$\frac{wd}{(1+\nu)ab^2\Delta T_{(b)}} = \frac{1}{(K+2)^2}\left[\left(\frac{r}{b}\right)^{K+2} - \left(\frac{K}{2}+1\right)\left(\frac{r}{b}\right)^2 + \frac{K}{2}\right] \tag{4.64}$$

$$\frac{12M_r}{Ed^2\alpha\Delta T_{(b)}} = \frac{1}{(K+2)}\left[\left(\frac{r}{b}\right)^K + \frac{1+\nu}{1-\nu}\right] \tag{4.65}$$

$$\frac{12M_\theta}{Ed^2\alpha\Delta T_{(b)}} = \frac{1}{(K+2)}\left[(K+1)\left(\frac{r}{b}\right)^K + \frac{1+\nu}{1-\nu}\right] \tag{4.66}$$

where $\Delta T_{(b)} = a_K b^K$.

For simply supported or free boundary conditions, the clamped plate results are modified by adding the effect of an edge moment $-M_r$ at $r = b$. This is another application of the *principle of superposition* mentioned in Section 4.4.3.

4.5.2. Bending of Circular Plates Due to Asymmetric Temperature Distributions

For this problem a convenient form of assumed temperature distribution is that suggested by Forray and Newman[43], viz.

$$\Delta T = \sum_{m=0}^{\infty}\sum_{k=0}^{\infty} C_{km} r^k \cos m\theta + \sum_{m=0}^{\infty}\sum_{k=0}^{\infty} S_{km} r^k \sin m\theta \tag{4.67}$$

Using eqn. (4.56) the approach is now very similar to that in the membrane problem of Section 3 which used eqn. (2.24). The deflexion w is expressed as the sum of the complete solution of $\nabla^4 w_c = 0$ and a particular solution of $\nabla^2 w_p = (1 + \nu)\alpha \Delta T/d$. The unknown coefficients in the complete solution are determined from the boundary conditions.

CHAPTER 5

Thermal Stresses in Beams and Circular Cylinders

5.1. Free Rectangular Beams

The analyses presented in Sections 2.5, 3.5 and 3.6, for thin rectangular plates are obviously also applicable to thin rectangular beams; with similar limitations on their validity. In this case the beam length, depth and width lie in the coordinate directions x, y and z respectively, in Fig. 2.1.

For temperature variations through the *depth* only, i.e. $T = T_{(y)}$, Section 2.5 gives an exact solution to the plane stress problem within the limitations of Saint-Venant's principle. In Sections 3.5 and 3.6 methods are given for satisfying precisely the traction-free boundary conditions at the ends of the beam; of which the methods in Section 3.6 permit problems of two-dimensional temperature distributions, i.e. $T = T_{(x,y)}$ to be considered.

The justification of using any arbitrary distribution, $T = T_{(x,y)}$ however, is not obvious since the plane stress hypothesis ideally requires that $\nabla^2(\alpha T) = F_{(z,t)}$, i.e. eqn. (2.9), although the analysis discussed in Section 2.2 does justify the use of $T = T_{(x,y)}$ for very thin members. To examine this point more closely reference is made to an analysis by Boley[14,15] which presents an *exact* two-dimensional thermoelastic solution for a rectangular beam subjected to temperature distributions of the form $T = T_{(x,y)}$. The solution is exact within the limitations of Saint-Venant's principle, i.e. the end tractions are only self-equilibrating and not generally zero. Therefore, the beam is assumed to be very long so that $a \gg b$ (Fig. 2.1) and the width, d, is sufficiently small so that the plane stress theory of Section 2.2 applies.

To obtain a solution to eqn. (2.24), viz.

$$\frac{\partial^4\phi}{\partial x^4} + 2\frac{\partial^4\phi}{\partial x^2\partial y^2} + \frac{\partial^4\phi}{\partial y^4} = -Ea\left(\frac{\partial^2 T}{\partial x^2} + \frac{\partial^2 T}{\partial y^2}\right) \tag{5.1}$$

the Airy stress function, ϕ, is expressed in the form

$$\phi = \phi_1 + \phi_2 + \phi_3 + \ldots \tag{5.2}$$

where each function ϕ_i satisfies the boundary conditions on $y = \pm b$, i.e.

$$\phi = \frac{\partial\phi}{\partial y} = 0 \tag{5.3}$$

Boley took the functions ϕ_i as follows:

$$\left.\begin{aligned} \frac{\partial^4\phi_1}{\partial y^4} &= -Ea\frac{\partial^2 T}{\partial y^2} \\ \frac{\partial^4\phi_2}{\partial y^4} &= -Ea\frac{\partial^2 T}{\partial x^2} - 2\frac{\partial^4\phi_1}{\partial x^2\partial y^2} \\ \frac{\partial^4\phi_i}{\partial y^4} &= -2\frac{\partial^4\phi_{i-1}}{\partial x^2\partial y^2} - \frac{\partial^4\phi_{i-2}}{\partial x^4} \text{ for } i = 3, 4 \end{aligned}\right\} \tag{5.4}$$

These functions can be shown to satisfy eqn. (5.1) and have the useful property that with T a known function of x and y each ϕ_i is readily determined from a knowledge of T and/or the preceding ϕ_i's. It is found that the expressions for the stress function ϕ, and also the stresses are given in terms of a series in which successive terms depend on successively higher derivatives of T with respect to x.

Thus if $T = yX_{(x)}$ or $T = X_{(x)}$ the first terms in the series for σ_{xx}, σ_{xy} and σ_{yy} are respectively proportional to the second, third and fourth derivatives of $X_{(x)}$, e.g. if $T = yX_{(x)}$

$$\frac{\sigma_{xx}}{Ea} = \frac{(3b^2 - 5y^2)y}{30}\frac{d^2X}{dx^2} + \left(\frac{y^5}{60} - \frac{y^3c^2}{30} + \frac{9yc^4}{700}\right)\frac{d^4X}{dx^4} + \ldots \tag{5.5}$$

The *elementary* solution for a temperature distribution which is linear or constant in y would predict zero stresses, irrespective of the form of $X_{(x)}$ so that the elementary solution is then not a good approximation to the above result if the second derivative of X is non-zero.

As an example, if $T = T_o\,(y/b)\,(x/a)^2$ then the only non-zero stress is σ_{xx}, which from (5.5) is given by

$$\frac{\sigma_{xx}}{E\alpha T_o} = \left(\frac{y}{b}\right)\left(\frac{3b^2 - 5y^2}{15a^2}\right) \tag{5.6}$$

This has a maximum value of

$$\left(\frac{\sigma_{xx}}{E\alpha T_o}\right)_{\max} = \pm\frac{2}{15}\left(\frac{b}{a}\right)^2 \text{ at } y = \pm b \tag{5.7}$$

and it follows that since $a \gg b$ for the above analysis to be valid then the stress σ_{xx} must be small. Since the elementary solution would have given $\sigma_{xx} = 0$ it is obvious that such a solution tends to be more accurate as the beam becomes more shallow.

The most important result of Boley's analysis is that the elementary solution of eqn. (2.38) gives the correct results for a long, thin rectangular beam, (except of course near the ends) if the temperature distribution $T = T_o + X_{(x)}\,Y_{(y)}$ is either uniform or linear in x. In other cases the elementary solution provides a good approximation if the temperature varies smoothly with x, because then the convergence of the series in the *exact* solution is rapid. Therefore, just as elementary beam theory gives satisfactory results when the bending moment is smoothly variable, so also does elementary one-dimensional thermal stress theory (Section 2.5) give satisfactory results even when the temperature varies along the length of the beam. This is especially true for beams of large length/depth ratio.

5.2. Free Beams of Arbitrary Cross-Section

Similar studies to those above for beams of arbitrary, solid, cross-section indicate that if the temperature distribution is uniform or linear in x, but otherwise arbitrary, the *exact* solution gives expressions for the axial stress, σ_{xx}, and for the Airy stress function, $\phi = \phi_{(y,z)}$, which satisfy the conditions of plane strain theory. However, the *exact* solution is rather tedious to undertake except in certain special cases, e.g. the rectangular beam, and recourse is usually made to more elementary methods of analysis. The most common of these is the theory of strength of materials based on the Euler–Bernouilli assumptions which are used in isothermal problems and have been found to be sufficiently accurate for many engineering problems. These assumptions are " that plane cross-sections remain plane and perpendicular to the beam axis during the given loading and/or heating " and " distortion of the cross-section due to Poisson ratio effects is negligible ". This Poisson ratio effect is included in the *exact* solution but its determination is quite tedious and its overall effect is probably small in general.(15)

Therefore in the analyses which follow in this Chapter for beams, based on the elementary theory, the conclusions of Section 5.1 and

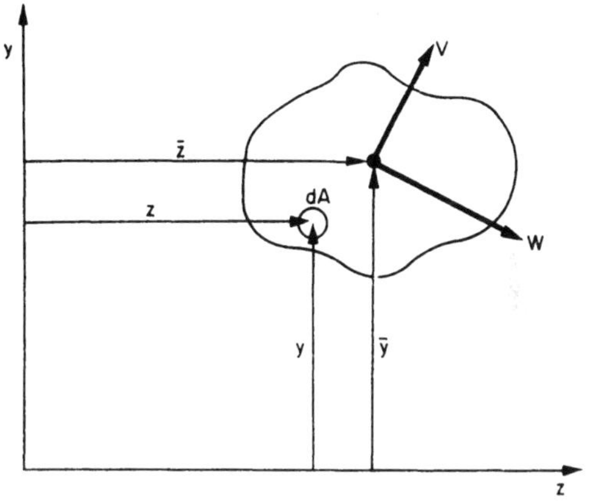

FIG. 5.1. Cross-section of beam.

above are implicitly incorporated, i.e. (1) the temperature distributions may be quite arbitrary ($T = T_{(x,y,z)}$) provided that the x-variations are smooth and (2) the Euler–Bernouilli assumptions are made.

From these assumptions it follows that the axial displacement, u, and the axial stress normal to the cross-section, σ_{xx}, may be written

$$u = C_1 + C_2 (y - \bar{y}) + C_3 (z - \bar{z}) \tag{5.8}$$

$$\sigma_{xx} = E \left(\frac{\partial u}{\partial x} - \alpha T \right) = E \left[C_1' + C_2' (y - \bar{\ }) + C_3' (z - \bar{z}) - \alpha T \right] \tag{5.9}$$

where compared with (1.14) it is seen that the strains normal to x have not been considered, i.e. $\nu = 0$. Here the parameters C_n are functions of x (primes denote differentiation with respect to x) and the distances y and z are measured from any convenient set of orthogonal axes in the cross-sectional planes as shown in Fig. 5.1. For the present problem the quantities C'_n may be determined from the requirements that the net axial force and moment on each cross-section should be zero, i.e. in the absence of external loads,

$$\int_A \sigma_{xx}\mathrm{d}A = \int_A \sigma_{xx}(y-\bar{y})\mathrm{d}A = \int_A \sigma_{xx}(z-\bar{z})\mathrm{d}A = 0 \qquad (5.10)$$

In eqns. (5.8)–(5.10) the parameters $\bar{y}$ and $\bar{z}$ denote the position of the elastic centroid of each cross-section measured relative to the assumed axes system. These parameters are defined later in eqn. (5.12).

If it is assumed that the beam material is linearly elastic at all temperatures; but that the elastic modulus (E) relating stress and strain may vary over the cross-section due to non-homogeneity or temperature, then the terms in (5.9) may be found, by using (5.10), to be

$$\left.\begin{aligned} C'_1 &= P/\overline{EA} \\ C'_2 &= [\overline{EI}_{yy} . M_{zz} - \overline{EI}_{yz} . M_{yy}]/[\overline{EI}_{yy} . \overline{EI}_{zz} - (\overline{EI}_{yz})^2] \\ C'_3 &= [\overline{EI}_{zz} . M_{yy} - \overline{EI}_{yz} . M_{zz}]/[\overline{EI}_{yy} . \overline{EI}_{zz} - (\overline{EI}_{yz})^2] \end{aligned}\right\} \qquad (5.11)$$

Positive moments about the $\overline{yy}$ and $\overline{zz}$ centroidal axes are those causing tensile stresses in the positive quadrant (where $y > \bar{y}$ and $z > \bar{z}$). P is positive when it is tensile and T is positive when measured above a datum level corresponding to a stress-free state.

The terms in eqn. (5.11) defining the elastic section properties of the cross-section are,

$$\left.\begin{aligned} \overline{EA} &= \oint E\mathrm{d}A \\ \bar{y} &= \oint Ey\mathrm{d}A/\oint E\mathrm{d}A \\ z &= \oint Ez\mathrm{d}A/\oint E\mathrm{d}A, \text{ and} \\ \overline{EI}_{yy} &= \oint Ez^2\mathrm{d}A - \bar{z}^2\oint E\mathrm{d}A \\ \overline{EI}_{zz} &= \oint Ey^2\mathrm{d}A - \bar{y}^2\oint E\mathrm{d}A \\ \overline{EI}_{yz} &= \oint Eyz\mathrm{d}A - \bar{y}\bar{z}\oint E\mathrm{d}A \end{aligned}\right\} \qquad (5.12)$$

and the *thermal end load* P and *thermal moments* M_{yy}, M_{zz} are given by

$$P = \oint E\alpha T \mathrm{d}A\,; \quad M_{yy} = \oint E\alpha T(z - \bar{z})\mathrm{d}A\,; \quad M_{zz} = \oint E\alpha T(y - \bar{y})\mathrm{d}A \tag{5.13}$$

where $\oint$ denotes integration over the cross-sectional area.

The results for $\bar{y}$ and $\bar{z}$ should be noted, viz. that the elastic centroid is the centroid of the "effective" elastic area, $\overline{EA}$, not the geometric area. The equations above, (5.12), apply to arbitrary axes y, z, but the bending stiffnesses about the elastic centroid principal axes can be written as

$$\overline{EI}_{vv}, \overline{EI}_{ww} = \frac{\overline{EI}_{yy} + \overline{EI}_{zz}}{2} \pm \sqrt{\left[\left(\frac{\overline{EI}_{yy} - \overline{EI}_{zz}}{2}\right)^2 + (\overline{EI}_{yz})^2\right]} \tag{5.14}$$

where $\overline{EI}_{vw} = \oint Evw\mathrm{d}A = 0$ defines the set of orthogonal elastic principal axes.

The above results are considerably simplified if the following conditions are realized.

I. If y and z axes coincide with principal centroidal axes v and w, i.e.

$$\sigma_{xx} = E[P/\overline{EA} + yM_{zz}/\overline{EI}_{zz} + zM_{yy}/\overline{EI}_{yy} - \alpha T] \tag{5.15}$$

II. If αT is symmetrical about a principal axis of symmetry of the elastic geometry, i.e. C_2' or C_3' equals zero.

The formulae given above apply to any unrestrained beam in which the variation of cross-section and temperature is continuous and smooth along its length, and they are valid for the entire beam except for small regions near the ends of the beam. To allow for end effects, analyses can be made similar to those for flat-plates—see Section 5.4.

5.3. Elementary Solutions for Free I-Beams

With multiweb structures of the type shown in Fig. 5.2 it is often permissible to idealize the structure to a set of I-beams, as shown in the figure, so as to reduce the complexity of the analysis. Thus each I-beam is considered quite separately and any discontinuities between adjacent I-beams are usually neglected. As a typical structural element the symmetrical I-beam has been considered often for both heat transfer and thermal stress analyses, (e.g. references 67, 130, 137, 147 and 148), as it represents a simple element of a built-up wing structure.

For such sections it is usually assumed that the flanges and web are sufficiently thin that temperature variations through their thickness may be neglected. This assumption results in a condition of plane stress in both flanges and web. The determination of the thermal stresses in a wide flanged I-beam having a symmetrical cross-section follows directly from the results of Section 5.2 and is particularly simple if the assumed temperature distribution is also

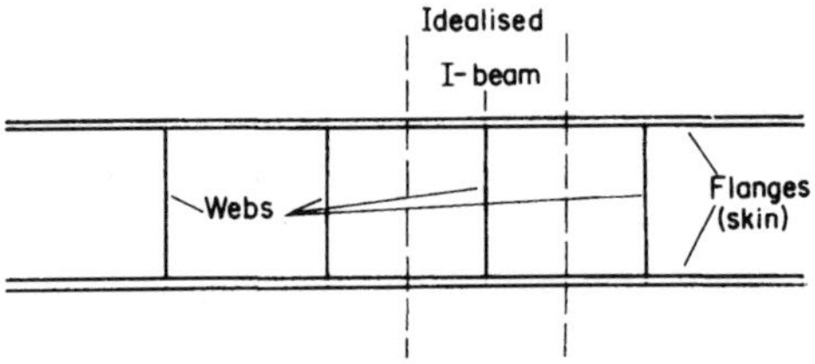

FIG. 5.2. Typical multiweb construction showing idealized I-beam element.

symmetrical. The governing equation for the longitudinal stress, σ_{xx}, then becomes,

$$\sigma_{xx} = E[P/\overline{EA} - \alpha T] \tag{5.16}$$

where $T = T_{(x,y,z)}$, provided the variations with x are smooth.

In the aircraft wing application it may often be assumed that the *convective heat transfer coefficient* and the *adiabatic wall temperature* (See Appendix) are uniform over the outer surfaces of both flanges. Because of the presence of the web the flange temperature at the web junction would be different from that at the flange edges; also, a temperature variation through the web depth would be expected. The precise form of the flange and web temperature distributions must be determined in each case for the particular environmental conditions (see references above), but it has been suggested that for preliminary design calculations parabolic distributions are reasonable approximations.[82]

Thus, during convective heating of the I-beam shown in Fig. 5.3, the web temperature distribution may be approximated by;

$$T_w = T_3 + (T_1 - T_3)\left(\frac{2y}{b_w}\right)^2, \quad T_1 > T_3 \tag{5.17}$$

and the flange distribution by

$$T_s = T_2 + (T_1 - T_2)\left(1 - \left|\frac{2z}{b_s}\right|\right)^2, \quad T_2 > T_1 \tag{5.18}$$

where T_1, T_2 and T_3 are functions of time and may also be functions of x, the axial coordinate of the beam. In reference 82 an approximate procedure is outlined by which the time-variation of T_1, T_2

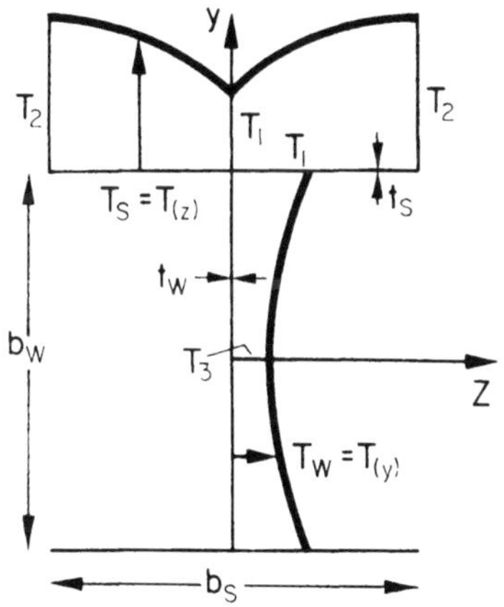

FIG. 5.3. Assumed temperature distribution in I-beam.

and T_3 may be determined for any arbitrary flight programme, and the resulting thermal stress distribution is also analysed. The following expression for σ_{xx} is obtained using eqns. (5.16)–(5.18)

$$\sigma_{xx} = \frac{E}{3}\left[\frac{E_s a_s A_s (T_1 + 2T_2) + E_w a_w A_w (T_1 + 2T_3)}{E_s A_s + E_w A_w} - 3aT\right] \tag{5.19}$$

In deriving this equation it is assumed that the flange and web materials are dissimilar but that their properties are invariant with temperature. A_s and A_w refer to the total flange (skin) and web cross-sectional areas respectively, i.e. $2t_s b_s$ and $t_w b_w$.

From eqn. (5.19) the maximum thermal stresses in the flange and web respectively are

$$(\sigma_{xx})_2 = E_s\,[E_w A_w\,\{\alpha_w(T_1 + 2T_3) - 3\alpha_s T_2\} - E_s A_s \alpha_s (T_2 - T_1)]/3\overline{EA} \qquad (5.20)$$

$$(\sigma_{xx})_3 = E_w\,[E_s A_s\,\{\alpha_s(T_1 + 2T_2) - 3\alpha_w T_3\} + E_w A_w \alpha_w (T_1 - T_3)]/3\overline{EA} \qquad (5.21)$$

where the suffixes 2, 3 refer to points 2, 3 in the cross-section and $\overline{EA} = E_s A_s + E_w A_w$.

5.4. The End Problem in Free I-Beams

The analyses of the previous sections have only satisfied the equilibrium conditions of zero net force and moment on each cross-section; so that point-wise satisfaction of traction-free boundary conditions at the ends of the beam is not achieved.

The thermal stress distributions near to the ends of wide flanged I-beams may be determined from analyses similar to those for thin rectangular plates or beams in Sections 3.5 and 3.6. Similar assumptions to those in Section 5.3 are usually made concerning the symmetry of structure and temperature distribution, and again, the flanges and web are assumed to be in a state of plane stress. The main complication to be overcome is concerned with the satisfaction of the boundary conditions at the flange-web joint. The conditions usually satisfied are those for compatibility of longitudinal strain and equilibrium of shear flow. Published analyses to date have only considered temperature distributions of the form $T = T_{(y,z)}$.

The approximate method developed by Isakson[76] consists of using an assumed form for the stress function ϕ in both flange and web, viz.

$$\phi = \phi_0 + a_1\phi_1 + a_2\phi_2 + \ldots \qquad (5.22)$$

where ϕ_0, ϕ_1, ϕ_2, . . ., etc., are the assumed functions and a_1, a_2, etc., are initially unknown multipliers. The forms of the series expressions for ϕ_w and ϕ_s must satisfy the various boundary conditions on the I-beam and the unknown a_n's may then be determined by a strain energy minimization process. The entire procedure is essentially the same as that in Section 3.5 (Method 1).

The method developed by Lempriere[100] is analogous to that of Section 3.6 (Method 4), but as presented it is only applicable to I-beams with no temperature variation across the width of the flanges. This method is developed further in reference 7 where more results are given and temperature variations across the flange are considered.

In reference 76 the " infinite theory " thermal stresses are first evaluated from (5.16) and then the corresponding stress functions $\phi_{Is} = f_1(z)$ and $\phi_{Iw} = f_2(y)$. These "first component " stresses at the free ends must then be cancelled by a " second component " stress distribution of opposite sign. The analysis for this " second component " distribution deals with a problem in isothermal elasticity, with the following expressions for ϕ_s and ϕ_w found to be the most convenient, viz.

$$\phi_s = \phi_{Is} + \sum_{m=0}^{p} \sum_{n=0}^{p} A_{mn} \left[\left(\frac{2z}{b_s} \right) - 1 \right]^2 \left[\frac{x^2}{l^2} - 1 \right]^2 \left(\frac{2z}{b_s} \right)^m \left(\frac{x}{l} \right)^{2n}$$

$$\phi_w = \phi_{Iw} + \sum_{m=0}^{p} \sum_{n=0}^{p} B_{mn} \left[\left(\frac{2y}{b_w} \right)^2 - 1 \right] \left[\frac{x^2}{l^2} - 1 \right]^2 \left(\frac{2y}{b_w} \right)^{2m} \left(\frac{x}{l} \right)^{2n} \tag{5.23}$$

where l is the beam semi-length.

The form of ϕ_s only applies in the range of positive z but because of symmetry the strain energy of the two halves of the flange must be equal so that only one side $(0 < z < b_s/2)$ needs to be considered in the energy minimization process. It should be observed that the forms of (5.23) automatically satisfy the required traction boundary conditions on the edges $x = \pm l$ and $z = \pm b_s/2$, viz.

$$\begin{aligned} &\text{at } z = b_s/2, \ \sigma_{zz} = \sigma_{xz} = 0 \\ &\quad x = \pm l, \ \sigma_{xx} = \sigma_{xx} \text{ for `` infinite '' theory, and } \sigma_{zx} = \sigma_{yx} = 0 \end{aligned} \tag{5.24}$$

Satisfaction of the boundary condition for equilibrium of shear flow at the flange-web joint leads to the following relations between the unknowns A_{mn} and B_{mn},

$$(2A_{on} - A_{1n})\ (b_w t_s / b_s t_w) = -\sum_{m=0}^{p} B_{mn}\ (n = 0, 1, 2, \ldots) \quad (5.25)$$

This equation provides $p + 1$ relations from which $p + 1$ of the unknowns can be expressed as linear functions of the remaining unknowns in the total set of $2(p + 1)^2$ unknowns.

The complementary strain energy in the beam is written in terms of the stress functions ϕ_s and ϕ_w and then minimized with respect to the various unknowns. This yields a set of $(p + 1)\ (2p + 1)$, linear, simultaneous equations in that number of unknowns. Knowing the various A_{mn} and B_{mn}, the corresponding stress distribution in the beam can be determined. This distribution coupled with the original "infinite theory" thermal stresses gives the required thermal stress distribution in an I-beam of finite length with all the important boundary conditions satisfied.

Parametric studies in references 7 and 100 show the effect of I-beam geometry on the end-problem in finite length beams. In particular, for a beam with $l/b_w > 3$ the longitudinal stress distribution at $x = 0$ is almost identical with that given by the "infinite theory". For shorter beams the "infinite theory" may greatly overestimate the maximum stress levels of σ_{xx}.

The maximum thermal shear stress resulting from the analyses referred to, occurs at a distance of about $b_w/2$ from the ends of the beam. Its maximum value may be more than 40 per cent of the maximum value of the longitudinal stress σ_{xx}.

Although the above analyses did not consider variations of temperature with x, there would appear to be no difficulty in extending Isakson's method(76) to allow for such a possibility. In this case the stress function forms ϕ_{Is} and ϕ_{Iw} only would be altered, to allow for x-wise variations in σ_{xx} as given by the elementary theory of (5.16), viz. $\phi_{Is} = F_1(x, z)$; $\phi_{Iw} = F_2(x, y)$. However, since the elementary theory only permits the existence of the stress σ_{xx}, then, in the expression for the strain energy, only the second derivatives of ϕ_{Is} and ϕ_{Iw}, with respect to z and y respectively, are considered.

5.5. Thermal Deflexions of Free Beams

From the analyses of Section 5.2 the axial displacement u is easily determined, i.e. from eqns. (5.8) and (5.9),

$$u = \int_0^x [C_1' + C_2'(y - \bar{y}) + C_3'(z - \bar{z})]\,\mathrm{d}x \tag{5.26}$$

where C_1', C_2' and C_3' are given by eqn. (5.11). Since the terms in C_2' and C_3' refer to curvatures of the beam the average axial displacement depends only upon C_1'. Thus

$$u_{av} = \int_0^x C_1'\,\mathrm{d}x \tag{5.27}$$

The deflexions v and w corresponding to the y and z directions also follow directly from the previous analyses, viz.

$$\frac{\mathrm{d}^2 v}{\mathrm{d}x^2} = -C_2' \; ; \; \frac{\mathrm{d}^2 w}{\mathrm{d}x^2} = -C_3' \tag{5.28}$$

For the special case of centroidal principal axes, eqns. (5.9) and (5.28) become,

$$\sigma_{xx} = E\left[\frac{P}{\overline{EA}} + \frac{M_{zz}\,.\,y}{\overline{EI}_{zz}} + \frac{M_{yy}\,.\,z}{\overline{EI}_{yy}} - \alpha T\right] \tag{5.29}$$

$$\frac{\mathrm{d}^2 v}{\mathrm{d}x^2} = -M_{zz}/\overline{EI}_{zz} \; ; \; \frac{\mathrm{d}^2 w}{\mathrm{d}x^2} = -M_{yy}/\overline{EI}_{yy} \tag{5.30}$$

where $$\overline{EI}_{yy} = \oint Ez^2\mathrm{d}A, \; \overline{EI}_{zz} = \oint Ey^2\mathrm{d}A \tag{5.31}$$

The complete similarity between these results and the corresponding ones of isothermal beam theory should be noted. It follows that in combined thermal and mechanical loading the above equations apply directly, provided that to the *thermal loads* P, M_{yy}, and M_{zz} are added the corresponding *mechanical loads* $\overline{P}$, $\overline{M}_{yy}$, and $\overline{M}_{zz}$.

Thermoelastic formulae for the analysis of beams making use of the concept of free energy (Section 1.6), are presented in a paper by Hemp(64). Energy methods such as these have particular application when beam deflexions are to be analysed.

5.6. Statically Indeterminate Beams (Externally Restrained)

In the previous sections equations are presented for statically determinate (free) beams, for stresses and deformations that are compatible with the internal requirement that plane sections remain plane and unrestrained externally. When there are external restraints on the beam additional stresses are caused so as to make the beam deflexions satisfy the given external boundary conditions. These additional stresses may be determined—once the thermal deflexions in the unrestrained (free) beam are known—by methods based on the conventional, isothermal theory of strength of materials.

No attempt is made here to formulate the technique or present detailed analyses for any of the many varied possible situations. It should be borne in mind however, that the problem is essentially still a thermal problem and the appropriate elastic section properties should be used, including the effects of temperature on the material properties, i.e. eqn. (5.12).

To illustrate the procedure consider a uniform beam under a temperature distribution $T = T_{(y,z)}$, so that the quantities C_1', C_2' and C_3' are independent of x. For the case of centroidal principal axes eqns. (5.29)–(5.31) apply. If this beam is externally restrained so that rotation of the cross-section about the z axis at x is prevented it follows from (5.26) that

$$C_2' y \int_0^x \mathrm{d}x = 0 \tag{5.32}$$

i.e. that $C_2' = 0$. By this is meant that the term C_2' must be cancelled to satisfy the specified restraint conditions. The resulting thermal stress distribution is then given by the modified (5.29), i.e.

$$\sigma_{xx} = E\left[\frac{P}{\overline{EA}} + \frac{M_{yy} \,.\, z}{\overline{EI}_{yy}} - \alpha T\right] \tag{5.33}$$

Obviously, if restraints act to prevent the average axial displacement and the rotations of the cross-section about both axes, all the terms C_1' to C_3' must be removed, and the stress state is simply given by

$$\sigma_{xx} = -\,E\alpha T_{(y,z)} \tag{5.34}$$

As another simple example consider the unstressed rectangular cantilever beam illustrated in Fig. 5.4, having a length L and depth $2b$. Given that there is a linear temperature variation through the depth, $T = \Delta T\ (y/2b)$, the problem is to evaluate the stress distribution if the free end of the beam is restrained transversely by a spring of linear stiffness K.

For the beam with no spring restraint the linear temperature distribution produces no thermal stress, but a curvature of $a\Delta T/2b$ is induced in the beam. Therefore, relative to the fixed end ($x = 0$) the free end of the beam at $x = L$ experiences a slope of $a\Delta TL/2b$ and a displacement of $a\Delta TL^2/4b$. The free end displacement is partially restrained by the spring which, in the equilibrium position,

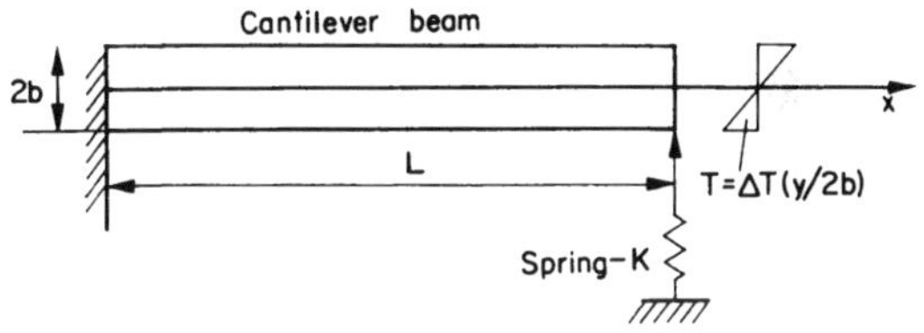

FIG. 5.4. Propped cantilever beam of rectangular cross-section.

experiences a force, P, corresponding to the spring displacement δ_s (where $P = K\delta_s$). Since the force P also acts on the cantilever beam, ordinary strength of materials theory yields the following relation between P and the corresponding end deflexion, δ_B of the beam, viz.

$$\delta_B = PL^3/3\,\overline{EI}_{zz} \tag{5.35}$$

Compatibility requires that

$$a\Delta TL^2/4b = \delta_B + \delta_s \tag{5.36}$$

therefore the force P is found to be,

$$P = \left[\frac{a\Delta TL^2}{4b}\right] \Big/ \left[\frac{L^3}{3\overline{EI}_{zz}} + \frac{1}{K}\right] \tag{5.37}$$

and for the beam the resultant stress distribution is given by

$$\sigma_{xx} = P(L - x)Ey/\overline{EI}_{zz} \tag{5.38}$$

For $K = \infty$, the conditions correspond to a rigidly-propped cantilever and the maximum bending stress is $\frac{3}{4}E\alpha\Delta T$.

If the free end of the beam is restrained by being " built-in " the *redundant force* at that end is a moment sufficient to negate the free thermal curvature, i.e.

$$M = \overline{EI}\,\alpha\Delta T/2b \tag{5.39}$$

The final situation is that in which bending stresses are set-up in a beam experiencing no net curvature. The maximum stress in this case is the same along the entire length of the beam, i.e. $\frac{1}{2}E\alpha\Delta T$.

5.7. Axisymmetric Thermal Stresses in Circular Cylinders

In Section 2.3 it is shown that solutions to the plane strain problem for circular cylinders may be found directly from known plane stress solutions by using the modified constants of eqn. (2.17). Also, for plane strain there is an additional non-zero stress component, σ_{zz} given by

$$\sigma_{zz} = \nu(\sigma_{rr} + \sigma_{\theta\theta}) - E\alpha T \tag{5.40}$$

If axial displacements are not prevented and the end faces of the cylinder are free of tractions a secondary solution must be obtained in the manner of Section 2.1, which is then added to the plane strain results. The only stress component affected by the secondary solution is σ_{zz} which has a net value in this case of

$$\sigma_{zz} = \sigma_{rr} + \sigma_{\theta\theta} \tag{5.41}$$

This result is valid for the entire length of the cylinder, whether it be solid or hollow, except near the ends since eqn. (5.41) only satisfies the condition of zero net force on the cross-section. Equation (5.41) may be written more simply as

$$\sigma_{zz} = \frac{E}{1-\nu}\left[(\alpha T)_{av} - \alpha T\right] \tag{5.42}$$

which is of course the plane strain equivalent of eqn. (5.16).

There are many published solutions for circular cylinders in the literature of which the first was by Duhamel[30]. References 159 and 163 treat the subject thoroughly, giving many other references, of particular note being the paper by Jaeger[77]. The various published solutions differ essentially in the form of heating environment considered; of these, two particular solutions will now be considered in more detail.

I. A hollow circular cylinder with temperatures T_a and T_b on the inner and outer surfaces respectively has the steady state temperature distribution,

$$T = T_b + (T_a - T_b) \log(b/r)/\log(b/a) \qquad (5.43)$$

Detailed results for this problem are given in references 8 and 163

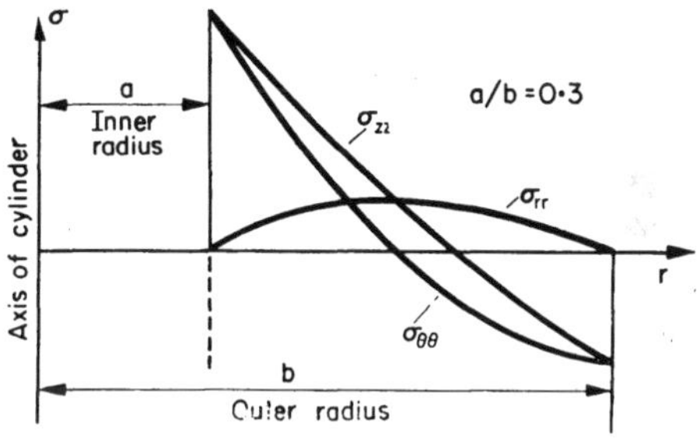

FIG. 5.5. Thermal stresses in hollow circular cylinder.

and are presented graphically in Fig. 5.5 for $a/b = 0{\cdot}3$. The tangential stress $\sigma_{\theta\theta}$ may be written as,

$$\sigma_{\theta\theta} = E\alpha(T_a - T_b) \times$$

$$\left[1 - \log(b/r) - \frac{a^2}{b^2 - a^2}\left(1 + \frac{b^2}{r^2}\right) \log(b/a)\right] \Big/ 2(1 - \nu) \log\left(\frac{b}{a}\right) \qquad (5.44)$$

and when $T_a > T_b$ $\sigma_{\theta\theta}$ has its maximum compressive value at $r = a$ and its maximum tensile value at $r = b$. Since $\sigma_{rr} = 0$ at $r = a$ and $r = b$, $\sigma_{zz} = \sigma_{\theta\theta}$ at these two extreme positions, from eqn. (5.41).

For a thin walled tube the following approximation can be made if h, the wall thickness, is much less than a. Then $(b - a)/a = h/a$

and,

$$\log(b/a) = (h/a) - \tfrac{1}{2}(h/a)^2 + \tfrac{1}{3}(h/a)^3 - \ldots \qquad (5.45)$$

Therefore,

$$\left.\begin{aligned}(\sigma_{\theta\theta})_a &= (\sigma_{zz})_a \doteqdot -E\alpha(T_a - T_b)\left(1 + \frac{h}{3a}\right)\Big/2(1-\nu) \\ (\sigma_{\theta\theta})_b &= (\sigma_{zz})_b \doteqdot E\alpha(T_a - T_b)\left(1 - \frac{h}{3a}\right)\Big/2(1-\nu)\end{aligned}\right\} \quad (5.46)$$

For the very thin wall the term $h/3a \ll 1$ and the stresses at $r = a$ and $r = b$ are equal but of opposite sign, i.e.

$$(\sigma_{\theta\theta})_a = -(\sigma_{\theta\theta})_b = -E\alpha(T_a - T_b)/2(1-\nu) \qquad (5.47)$$

It is seen that the cylinder wall thickness does not enter directly into this equation, but it can be inferred that since there would probably be a greater temperature difference in the case of a thicker wall, the thermal stresses would then also be greater. Similarly it should be noted that for a hollow cylinder of very small bore, i.e. $a/b \rightarrow 0$, eqn. (5.44) reduces to

$$\left.\begin{aligned}(\sigma_{\theta\theta})_a &= -E\alpha(T_a - T_b)/(1-\nu) \\ (\sigma_{\theta\theta})_b &= 0\end{aligned}\right\} \qquad (5.48)$$

Comparison of these results with (5.46) shows that the maximum stress is twice as great in the " solid " cylinder as for the thin walled tube even for the same temperature difference.

II. A solid cylinder has a uniform heat generation rate H at all interior points and a prescribed circumferential surface temperature, T_s. The steady state temperature distribution is easily shown to be

$$T = H(b^2 - r^2)/4k + T_s \qquad (5.49)$$

satisfying the equation, (A.8 of the Appendix)

$$k\left(\frac{d^2T}{dr^2} + \frac{1}{r}\frac{dT}{dr}\right) + H = 0 \qquad (5.50)$$

The corresponding thermal stresses are,

$$\left.\begin{aligned}\sigma_{rr} &= \bar{H}(r^2 - b^2)\\ \sigma_{\theta\theta} &= \bar{H}(3r^2 - b^2)\\ \sigma_{zz} &= \bar{H}(4r^2 - b^2)\\ \text{where} \qquad \bar{H} &= E\alpha H/16k(1 - \nu)\end{aligned}\right\} \tag{5.51}$$

These results have direct application to thermal stress analyses of power-producing elements typical of those used in nuclear reactors. For more detailed considerations of this special field of problems the reader should study references 53, 160 and 161.

CHAPTER 6

Thermal Stresses in Shells

6.1. Introduction

A shell is a material body which lies in a region bounded by two surfaces which are at equal perpendicular distances ($d/2$) on either side of a middle surface. The coordinates of the middle surface and the thickness completely define the shell.

The derivation of the governing equations for the deformations and stresses in shells is very similar to that for flat plates, in that the shell is assumed to be thin so that a state of two-dimensional plane stress exists. Obvious differences with thin plate bending theory do occur, e.g. in the form of the strain–displacement relations and in the terms appearing in the equations of equilibrium. Thus, unlike plate theory, there is no real measure of agreement, amongst the various published papers and texts, on the form the complete set of shell equations should have. Because of this, it would be inappropriate here to attempt to derive in detail the thermoelastic equations for shells of arbitrary shape, since this would, logically, necessitate a critical assessment of the entire subject of isothermal shell theories. Such theories are already well developed in references 37, 129 and 164. Therefore the emphasis in this chapter will be on those equations which are available for thermoelastic analyses of shells.

It should be noted that from analyses for shells of arbitrary shape under arbitrary temperature distributions considerable simplification occurs when dealing with shells of revolution, particularly for axisymmetrical loading and/or heating conditions. Amongst such shells are toroids, ogives, cones, spheres and cylinders, of which the last has received the most attention in the literature.

In references 60 and 61 Heldenfels presents a numerical method for the stress analysis of stiffened shell structures of arbitrary cross-section under non-uniform temperature distributions. Whilst not a rigorous method of solution it is sufficiently accurate and particularly useful for complicated structures that cannot be satisfactorily analyzed by simplified methods of analysis. Such procedures and others are also discussed by Broglio and Santini[16]; they are not considered further here.

6.2. Shells of Arbitrary Shape

General equations for the deformation of thin shells of arbitrary, variable, curvature can be found in the works of (amongst others) Love[103] and Novozhilov[129]. The extension of these results to the thermoelastic problem only requires the use of the appropriate strain–stress relations. This is recognized by Yao[173] who has extended the isothermal analysis of Novozhilov[129].

The expressions for the force and moment resultants have the same form as those for flat plates (4.8) and (4.9), i.e.

$$\left.\begin{aligned} N_1 &= \frac{Ed}{1-\nu^2}(e_1+\nu e_2) - \frac{N_T}{1-\nu} \\ N_2 &= \frac{Ed}{1-\nu^2}(e_2+\nu e_1) - \frac{N_T}{1-\nu} \\ N_{12} &= \frac{Ed}{2(1+\nu)}\cdot\omega \end{aligned}\right\} \tag{6.1}$$

$$\left.\begin{aligned} M_1 &= D(\kappa_1+\nu\kappa_2) - \frac{M_T}{1-\nu} \\ M_2 &= D(\kappa_2+\nu\kappa_1) - \frac{M_T}{1-\nu} \\ M_{12} &= D(1-\nu)\tau, \text{ where } D = Ed^3/12(1-\nu^2) \end{aligned}\right\} \tag{6.2}$$

The relations between the curvature changes and displacements are given by

$$\left.\begin{aligned} \kappa_1 &= -\frac{1}{A_1}\frac{\partial}{\partial\alpha_1}\left(\frac{1}{A_1}\frac{\partial w}{\partial\alpha_1}\right) - \frac{1}{A_1A_2}\frac{\partial A_2}{\partial\alpha_2}\frac{1}{A_2}\frac{\partial w}{\partial\alpha_2} \\ \kappa_2 &= -\frac{1}{A_2}\frac{\partial}{\partial\alpha_2}\left(\frac{1}{A_2}\frac{\partial w}{\partial\alpha_2}\right) - \frac{1}{A_1A_2}\frac{\partial A_2}{\partial\alpha_1}\frac{1}{A_1}\frac{\partial w}{\partial\alpha_1} \\ \tau &= \frac{-1}{A_1A_2}\left[\frac{\partial^2 w}{\partial\alpha_1\partial\alpha_2} - \frac{1}{A_1}\frac{\partial A_1}{\partial\alpha_2}\frac{\partial w}{\partial\alpha_1} - \frac{1}{A_2}\frac{\partial A_2}{\partial\alpha_1}\frac{\partial w}{\partial\alpha_2}\right] \end{aligned}\right\} \tag{6.3}$$

where A_1 and A_2 are the Lamé parameters associated with the particular orthogonal coordinate system used (α_1, α_2), and w is the displacement normal to the middle surface of the shell. In (6.3) the terms dependent upon the tangential displacement components are neglected.

The force-resultants are replaced in the subsequent analysis by introducing a stress function ϕ_T, to which the quantities N_1, N_2 and N_{12} can be related, and which enables two of the six equations of equilibrium to be automatically satisfied (neglecting certain terms) —cf. with eqns. (4.12) and (4.13). The remaining equations of equilibrium can be reduced by suitable manipulation to give the first governing equation, cf. eqns. (4.16)–(4.20), viz.

$$D\Delta\Delta w + \Delta M_T/(1-\nu) - B\phi_T = Z \tag{6.4}$$

where
$$\left.\begin{aligned} \Delta &= \frac{1}{A_1A_2}\left[\frac{\partial}{\partial\alpha_1}\left(\frac{A_2}{A_1}\frac{\partial}{\partial\alpha_1}\right) + \frac{\partial}{\partial\alpha_2}\left(\frac{A_1}{A_2}\frac{\partial}{\partial\alpha_2}\right)\right] \\ B &= \frac{1}{A_1A_2}\left[\frac{\partial}{\partial\alpha_1}\left(\frac{1}{R_2}\frac{A_2}{A_1}\frac{\partial}{\partial\alpha_1}\right) + \frac{\partial}{\partial\alpha_2}\left(\frac{1}{R_1}\frac{A_1}{A_2}\frac{\partial}{\partial\alpha_2}\right)\right] \end{aligned}\right\} \tag{6.5}$$

Z = distributed transverse pressure on shell, and R_1, R_2 are the principal radii of curvature of the shell.

The second governing equation is obtained from the equation of compatibility; which, since the strains e can be related to ϕ_T by means of (6.1), results in the following equation for ϕ_T, viz.

$$\Delta\Delta\phi_T + EdBw + \Delta N_T = 0 \tag{6.6}$$

when certain higher order terms are neglected.

Equations (6.4) and (6.6) can be used to derive approximate solutions for the *small deflexion* behaviour of thin shells subjected to transverse loading Z and a temperature distribution, T. It is seen that these equations are coupled through ϕ_T and in this respect the results are different from those for small deflexion behaviour of flat plates and more similar to those for the *large deflexion* behaviour of flat plates [eqns. (4.25) and (4.26)].

It would appear that by operating on eqns. (6.4) and (6.6) by the differential operators $\Delta\Delta$ and B respectively, the terms in ϕ_T can be eliminated, thus,

$$D\Delta\Delta\Delta\Delta w + \Delta\Delta\Delta M_T/(1-\nu) + EdBBw + B\Delta N_T = \Delta\Delta Z \tag{6.7}$$

An alternative study of this problem is given in a paper by Hemp[65] which presents methods of analysis for the *large deflexion* behaviour of thin shells with a constant temperature through the thickness. Some special cases are examined including the problem of thermal buckling.

By extending this work[65] to allow for temperature variations through the thickness, the following equations are obtained for *large local deformations* of a thin heated shell subjected also to a transverse pressure, Z. It is assumed that the geometrical parameters A_1, A_2, R_1 and R_2 are sensibly constant in the region of interest near the point $(\bar{\alpha}_1, \bar{\alpha}_2)$, which must not be too close to a singular point of the coordinate system. For convenience the notation x, y is used instead of $A_1\alpha_1$, $A_2\alpha_2$.

$$D\nabla^4 w + \nabla^2 M_T/(1-\nu) - \left(\frac{1}{R_1} + \underline{\frac{\partial^2 w}{\partial x^2}}\right)\frac{\partial^2 \phi_T}{\partial y^2} - \left(\frac{1}{R_2} + \underline{\frac{\partial^2 w}{\partial y^2}}\right)\frac{\partial^2 \phi_T}{\partial x^2} + \underline{2\frac{\partial^2 w}{\partial x \partial y}\cdot\frac{\partial^2 \phi_T}{\partial x \partial y}} = Z \tag{6.8}$$

$$\nabla^4 \phi_T + Ed\left[\frac{1}{R_1}\frac{\partial^2 w}{\partial y^2} + \frac{1}{R_2}\frac{\partial^2 w}{\partial x^2} + \underline{\frac{\partial^2 w}{\partial x^2}\frac{\partial^2 w}{\partial y^2}} - \underline{\left(\frac{\partial^2 w}{\partial x \partial y}\right)^2}\right] + \nabla^2 N_T = 0 \tag{6.9}$$

where
$$\nabla^2 = \frac{\partial^2}{\partial x^2} + \frac{\partial^2}{\partial y^2}$$

The difference between these equations and the small deflexion equations (6.4) and (6.6) is mainly in the non-linear terms which are underlined. The agreement between these equations, when R_1 and R_2 are infinite, and their counterparts for flat plates, (4.25) and (4.26), is immediately obvious. It should be noted that eqns. (6.8) and (6.9) have been presented previously by Marguerre[109].

The analysis of thermoelastic problems of shells of arbitrary curvature is obviously most difficult and apparently there are no published solutions to such problems. Even with the " simple " equations (6.8) and (6.9) the difficulties are considerable, because of the inherent non-linearities present. Iterative methods of solution starting from the linear, uncoupled forms of (6.8) and (6.9) are possible but tedious. Alternatively, approximate methods of analysis using an assumed form for w, solving (6.9) for ϕ_T and substituting both w and ϕ_T into an energy expression equivalent to (6.8) might be more practicable. Since the solution of coupled, non-linear equations is a problem in thermal buckling and post-buckling further discussion on this topic is deferred until Chapter 7.

It should be pointed out that, irrespective of the form of the cross-section, a closed thin shell having a linear gradient of temperature across its thickness with uniform surface temperatures T_1, T_2 experiences bending moments necessary to annul the free thermal curvature as given by eqn. (4.39). Thus the maximum bending stresses are

$$(\sigma_{11})_{\max} = (\sigma_{22})_{\max} = \pm\, E\alpha(T_2 - T_1)/2(1 - \nu) \qquad (6.10)$$

at distances remote from any free edges. At such edges the condition of zero traction normal to the free edge induces a local irregularity into the uniform stress distribution (eqn. (6.10)). In the case of a circular cylindrical shell the corresponding maximum circumferential stress is 25 per cent greater than the value in (6.10).

6.3. Shells of Revolution with a Meridian of Arbitrary Shape

Thermoelastic equations for shells of revolution with a meridian of arbitrary shape may be derived directly from Flugge's isothermal equations[37] which consist of six equations of equilibrium (one of which is an identity), and eight expressions relating the force and moment resultants to the displacement components of the middle surface (u, v and w). Thus the thirteen independent equations are sufficient to determine the thirteen unknowns (N_1, N_2, N_{12}, N_{21}, M_1, M_2, M_{12}, M_{21}, Q_1, Q_2, u, v, w). In practice[129] it is usually assumed that the two shears N_{12}, N_{21} are numerically equal (to N_{12}) and also the two twisting moments M_{12}, M_{21} (to M_{12}) so that there are only eleven unknowns to be found from eleven useful equations. The two transverse shear forces Q_1, Q_2 have a similar significance to the corresponding terms, Q_x, Q_y, in Section 4.1 (eqn. (4.17)). Whether or not the last assumption is made (it was in Section 6.2) there is sufficient information to solve a given problem and a typical procedure would be to eliminate all the unknowns except u, v, and w, which would then appear in a set of three simultaneous differential equations, containing terms dependent upon the temperature distribution and/or the external loading on the shell. By successive differentiation of these three equations and elimination of less significant terms it should be possible to reduce the analysis finally to one differential equation in w, the transverse displacement of the middle surface. Such an equation could of course have been derived from the more general equation of Section 6.2, viz. (6.7). (The procedure described above is that followed, for instance, in deriving Donnell's eighth-order equation[28] in w for circular cylindrical shells.)

When the loads and temperatures are axisymmetric there is considerable simplification and it is useful to pursue the development of the governing equations in some detail.

The general configuration of the shell is shown in Fig. 6.1 and the positive directions of the non-zero force and moment resultants are shown in Fig. 6.2. Because of symmetry many of the resultants vanish identically, e.g. N_{12}, N_{21}, M_{12}, M_{21} and Q_2, and only two external load components need to be considered, p_r and p_ψ, (p_r is transverse to the shell surface and p_ψ is tangential to the shell meridian). For any element bounded by the angles $d\theta$ and $d\psi$ the following equations of equilibrium are obtained,[37,164]

$$\left.\begin{aligned}
&\frac{d}{d\psi}(N_1R_2\sin\psi) - N_2R_1\cos\psi - Q_1R_2\sin\psi + R_1R_2\sin\psi\,.\,p_\psi = 0\\
&\frac{d}{d\psi}(Q_1R_2\sin\psi) + N_1R_2\sin\psi + N_2R_1\sin\psi + R_1R_2\sin\psi\,.\,p_r = 0\\
&\frac{d}{d\psi}(M_1R_2\sin\psi) - M_2R_1\cos\psi - Q_1R_1R_2\sin\psi = 0
\end{aligned}\right\} \tag{6.11}$$

There are five unknown quantities in these three equations, but the number may be reduced to three by expressing the forces N_1, N_2 and the moments M_1, M_2 in terms of the components of displacement u, w. These expressions are given approximately by[37,164]

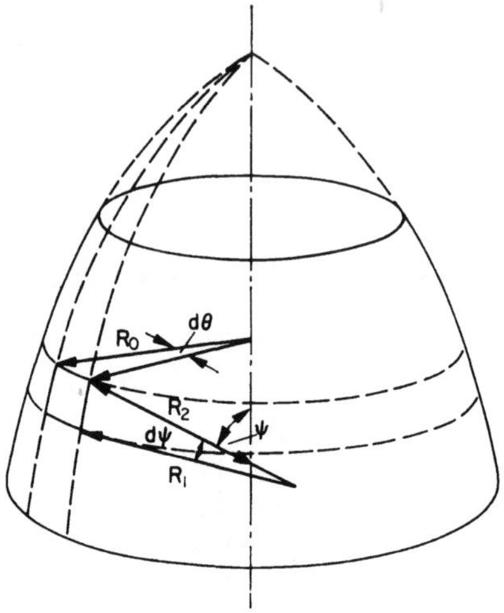

FIG. 6.1. General configuration of an axisymmetric shell.

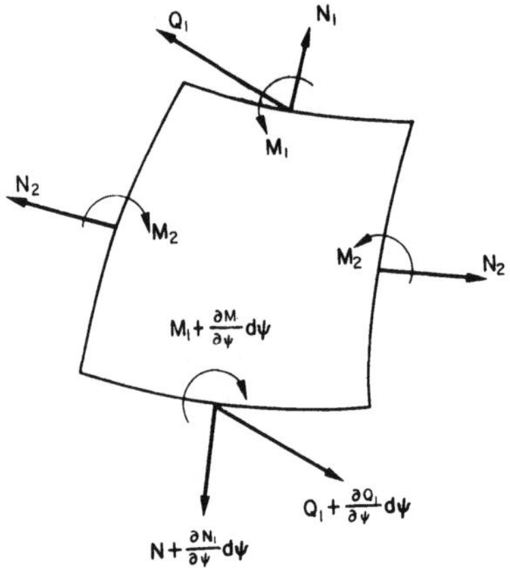

FIG. 6.2. Force and moment resultants on shell element.

$$\left.\begin{aligned}
N_1 &= \frac{Ed}{1-\nu^2}\left[\frac{1}{R_1}(u'-w)+\frac{\nu}{R_2}(u\cot\psi-w)\right]-\frac{N_T}{1-\nu}\\
N_2 &= \frac{Ed}{1-\nu^2}\left[\frac{1}{R_2}(u\cot\psi-w)+\frac{\nu}{R_1}(u'-w)\right]-\frac{N_T}{1-\nu}\\
M_1 &= -D\left[\frac{1}{R_1}\left(\frac{u+w'}{R_1}\right)'+\frac{\nu}{R_2}\left(\frac{u+w'}{R_1}\right)\cot\psi\right]-\frac{M_T}{1-\nu}\\
M_2 &= -D\left[\frac{1}{R_2}\left(\frac{u+w'}{R_1}\right)\cot\psi+\frac{\nu}{R_1}\left(\frac{u+w'}{R_1}\right)'\right]-\frac{M_T}{1-\nu}
\end{aligned}\right\} \quad (6.12)$$

where the primes denote differentiation with respect to ψ. Substituting these expressions into eqn. (6.11) yields three equations in only three unknown quantities u, w, and Q_1. These three equations may be reduced to two, by eliminating Q_1, with unknowns u and w, which may then be solved for the particular loading and temperature distributions, shell geometry and edge boundary conditions.

However, the equations may be written in terms of different variables in a form more amenable to solution. These variables are $V = (u + w')/R_1$ and $U = R_2Q_1$. To ease the introduction of these new variables into the analysis the first equation of (6.11), which defines the equilibrium in the meridional direction of the in-plane forces on an element, is replaced by one defining the overall equilibrium of the shell above the parallel circle of radius R_0 (Fig. 6.1). If, therefore, the net vertical resultant of the external loadings p_ψ and p_r is denoted by F, this equation gives (see Fig. 6.3)

$$2\pi R_0 \sin\psi\, N_1 + 2\pi R_0 \cos\psi\, Q_1 + F = 0 \qquad (6.13)$$

from which

$$N_1 = -U\cot\psi/R_2 - F/2\pi R_2 \sin^2\psi \qquad (6.14)$$

The second equation of (6.11) may be written alternatively as

$$N_2R_1\sin\psi = -N_1R_2\sin\psi - \frac{d}{d\psi}(Q_1R_2\sin\psi) - R_1R_2\sin\psi\,.\,p_r$$

and substituting (6.14) yields

$$N_2 = -\frac{1}{R_1}\frac{dU}{d\psi} - p_r\,.\,R_2 + F/2\pi R_1\sin^2\psi \qquad (6.15)$$

From the equations for N_1, N_2, i.e. eqn. (6.12), the following expressions are obtained,

$$u' - w = \frac{R_1}{Ed}\left[N_1 - \nu N_2 + N_T\right] \tag{6.16}$$

$$u \cot \psi - w = \frac{R_2}{Ed}\left[N_2 - \nu N_1 + N_T\right] \tag{6.17}$$

By suitably operating on the left-hand sides of these two equations, and then combining them, the following equation may be obtained,

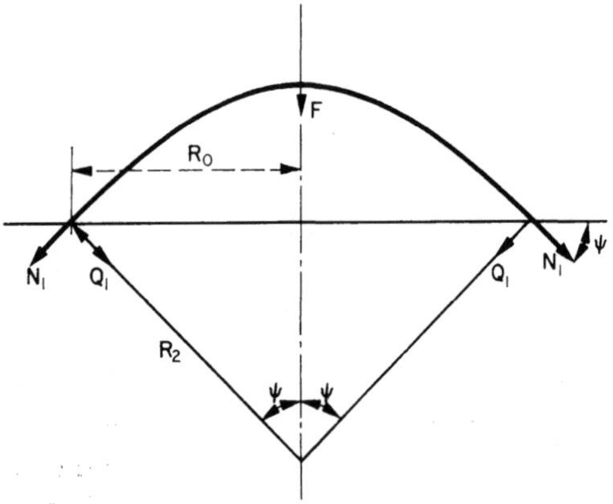

FIG. 6.3. Overall equilibrium of shell.

$$R_1 V = \frac{\cot \psi}{Ed}\left[(R_1 + \nu R_2)N_1 - (R_2 + \nu R_1)N_2 + N_T(R_1 - R_2)\right] - \frac{d}{d\psi}\left[\frac{R_2}{Ed}(N_2 - \nu N_1 + N_T)\right] \tag{6.18}$$

By substituting expressions (6.14) and (6.15) into (6.18) a relation between U and V is obtained in terms of the quantities F, p_r and N_T. The following geometrical relation is used in this substitution

$$R_2' = (R_1 - R_2)\cot \psi \tag{6.19}$$

The second equation for U and V is obtained from the third of eqn. (6.11) by substituting the expressions for M_1 and M_2 in (6.12). These expressions may be written in terms of V directly.

In the absence of external loads the governing equations for U and V are,[164]

$$U'' + \left[\frac{R_1}{R_2}\left(\frac{R_2}{R_1}\right)' + \cot\psi - \frac{1}{d}(d)\right]U'$$
$$-\left[\frac{R_1^2}{R_2^2}\cot^2\psi - \nu\frac{R_1}{R_2} - \frac{\nu R_1}{dR_2}\cot\psi(d)'\right]U = \frac{R_1^2}{R_2}EdV + R_1 d\left(\frac{N_T}{d}\right)' \tag{6.20}$$

and

$$V'' + \left[\frac{R_1}{R_2}\left(\frac{R_2}{R_1}\right)' + \cot\psi + \frac{3}{d}(d)'\right]V'$$
$$-\left[\frac{R_1^2}{R_2^2}\cot^2\psi + \nu\frac{R_1}{R_2} - \frac{3\nu R_1}{dR_2}\cot\psi\,(d)'\right]V = \frac{-R_1^2}{DR_2}U - \frac{R_1 M_T'}{D(1-\nu)} \tag{6.21}$$

where N_T and M_T are implicit functions of the shell thickness, d, and where generality has been retained in allowing R_1, R_2, d, N_T and M_T to be functions of ψ. It is seen that (6.20) and (6.21) are non-homogeneous simultaneous differential equations of the second order, for which closed form analytic solutions do not, in general, appear to be possible. It is probable that numerical integration offers the only method of solution. Obviously, for shells of constant thickness the terms in $(d)'$ are eliminated and the analysis is somewhat simpler.[37,164]

Once the quantities U and V are determined the various stresses and deformations can be found using previously derived relations. Thus M_1 and M_2, which can be expressed in terms of V, are easily evaluated; N_1 and N_2 follow from eqns. (6.14) and (6.15); u and w are then found by solving eqns. (6.16) and 6.17). The displacement $\bar{w}$ along the outward normal to the axis of symmetry is given by

$$\bar{w} = u\cos\psi - w\sin\psi = \sin\psi\,(u\cot\psi - w) \tag{6.22}$$

which may be determined directly from eqn. (6.17) as

$$\bar{w} = R_2\sin\psi\,[N_2 - \nu N_1 + N_T]/Ed \tag{6.23}$$

6.4. Shells of Revolution of Constant Meridional Curvature

From the analysis of the preceding section it is clear that considerable simplification occurs for that class of shells which have constant meridional curvature, R_1, i.e. toroids, ogives, spheres, cones and cylinders (see Fig. 6.4). For shells, of constant thickness, d, and deforming axisymmetrically, eqns. (6.20) and (6.21) can be written,

$$\left.\begin{aligned} L(U) + \frac{\nu R_1}{R_2} U &= \frac{R_1^2}{R_2} Ed\,.\,V + R_1 N_T' \\ L(V) - \frac{\nu R_1}{R_2} V &= -\frac{R_1^2 U}{DR_2} - \frac{R_1 M_T'}{D(1-\nu)} \end{aligned}\right\} \tag{6.24}$$

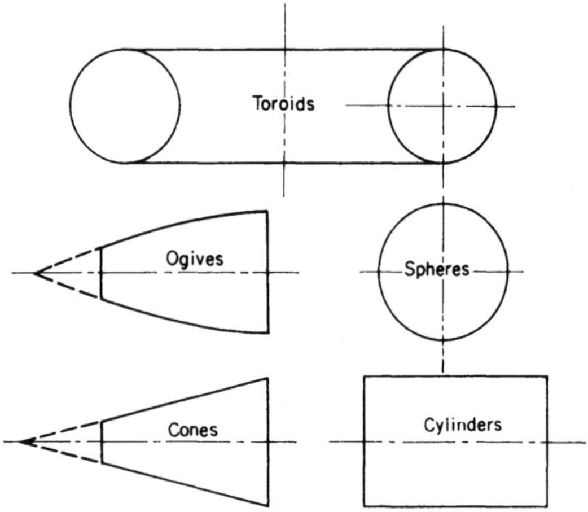

FIG. 6.4. Shells of constant meridional curvature.

where

$$L(\ ) = (\ )'' + \frac{R_1}{R_2}\cot\psi\,(\ )' - \frac{R_1^2}{R_2^2}\cot^2\psi\,(\ ) \tag{6.25}$$

if (6.19) is used. By defining $L_1(\ ) = R_2/R_1^2 L(\ )$ the two equations (6.24) can, after suitable manipulation, be replaced by two fourth-order equations for U and V independently. Thus the equation obtained in U is,

$$L_1 L_1(U) + 4\beta_1^4 U = \frac{1}{R_1} L_1(R_2 N_T') - \frac{\nu R_2}{R_1^2} N_T' - \frac{R_2 Ed M_T'}{R_1 D(1-\nu)} \tag{6.26}$$

where $$4\beta_1^4 = \frac{Ed}{D} - \frac{\nu^2}{R_1^2} \simeq \frac{Ed}{D} = \frac{12(1-\nu^2)}{d^2} \tag{6.27}$$

Having obtained the complete solution for U, the function V is determined from the second eqn. (6.24).

The complete solution to (6.26) consists of two parts, viz.

(1) the complementary function with four constants of integration and

(2) the particular integral which satisfies the given temperature " loadings " on the right-hand side of (6.26).

These two parts, the homogeneous and non-homogeneous solutions involve different types of functions, according to the type of shell, which are summarized in Tables 6.1 and 6.2 below. The given temperature " loadings " must be expressed in terms of the same functions as U and V in Table 6.2 only with known coefficients. The unknown coefficients in the series for U and V are then determined by equating corresponding terms.

TABLE 6.1.* SUMMARY OF COMPLEMENTARY FUNCTIONS

Type of Shell	Type of Functions
Ogival	Series solution } Special types of
Spherical	Series solution } hypergeometric functions
Conical	Thompson functions; ber, bei, ker, kei
Cylindrical	Trigonometric and hyperbolic functions

TABLE 6.2.* SUMMARY OF FUNCTIONS FOR PARTICULAR INTEGRALS

Type of Shell	Type of Functions
Spherical	Associated Legendre functions of the first kind, order one, $P_n^{(1)}$ (cos ψ)
Conical	Bessel function of the first kind, and order two, $J_2(x)$
Cylindrical	Trigonometric functions (cos x or sin x)

* These summaries were presented in an unpublished paper by J. S. Przemieniecki " Thermal Stresses in Shells ", Feb. 1961, which dealt with the subject matter of this chapter in some detail.

An alternative approximate method of analyzing the stresses in slender axisymmetric shells may be used, based on the method of asymptotic integration. This is discussed fully in reference 164. Analyses using the above procedure as applied to conical shells are given in reference 75, and other analyses are presented in references 151, 154 and 159. Thermal stresses in spherical shells are considered in reference 164, and in ogival shells in reference 25. The majority of shell analyses have been for cylindrical shells, which will be considered separately in the next section.

6.5. Circular Cylindrical Shells

6.5.1. Arbitrary Temperature Distributions

For a circular cylindrical shell the general procedure outlined in Section 6.3 may be followed, leading at an intermediate stage to only three equations of equilibrium and, at a later stage, to three equations in only three unknowns, e.g. the displacement components, u, v and w.

The three equations of equilibrium in the axial, circumferential and radial directions are, using the sign convention of reference 164, respectively,

$$\left.\begin{aligned} \frac{\partial N_1}{\partial x} + \frac{\partial N_{21}}{\partial y} + X &= 0 \\ \frac{\partial N_2}{\partial y} + \frac{\partial N_{12}}{\partial x} + \frac{1}{R}\frac{\partial M_{12}}{\partial x} - \frac{1}{R}\frac{\partial M_2}{\partial y} + Y &= 0 \\ \frac{N_2}{R} + \frac{\partial^2 M_{21}}{\partial x \partial y} + \frac{\partial^2 M_1}{\partial x^2} - \frac{\partial^2 M_{12}}{\partial x \partial y} + \frac{\partial^2 M_2}{\partial y^2} + Z &= 0 \end{aligned}\right\} \quad (6.28)$$

where X, Y, Z are the external loadings per unit area in the x, y and z directions, and R is the shell radius. The expressions for the force and moment resultants in terms of the strains are similar in form to those in (6.1) and (6.2); but it is the relations between the strains and the displacement components which must be decided before the three equations in the three unknowns u, v and w, can be obtained.

There is considerable difference of opinion in the literature over the precise form the aforementioned relations should have, resulting in wide variations in the three displacement equations. Flugge[37] discusses some of the approximations that can be made in formulating these relations and presents alternative sets of the displacement equations. To generalize, however, the following expressions indicate the form the equations should have; the forms of the operators A_{ij} depend upon the particular strain–displacement relations used;

$$\left.\begin{aligned} A_{uu}\,.\,u + A_{uv}\,.\,v + A_{uw}\,.\,w + \overline{X}\,(1-\nu^2)/Ed &= 0 \\ A_{vu}\,.\,u + A_{vv}\,.\,v + A_{vw}\,.\,w + \overline{Y}\,(1-\nu^2)/Ed &= 0 \\ A_{wu}\,.\,u + A_{wv}\,.\,v + A_{ww}\,.\,w + \overline{Z}\,(1-\nu^2)/Ed &= 0 \end{aligned}\right\} \quad (6.29)$$

The general form of A_{ij} is $(B_{ij} + C_{ij}d^2/12R^2)$ where the operators B_{ij} and C_{ij} are differential expressions e.g. $\partial^2/\partial x^2$ or $\partial^3/\partial x^2\partial y$, etc. The symbols $\overline{X}$, $\overline{Y}$ and $\overline{Z}$ now refer to the combined mechanical and thermal loading on the shell; the thermal loading being dependent upon the parameters N_T and M_T, viz.

$$\left.\begin{aligned} \overline{X} &= X - \frac{1}{1-\nu}\frac{\partial N_T}{\partial x} \\ \overline{Y} &= Y - \frac{1}{1-\nu}\frac{\partial N_T}{\partial y} + \frac{1}{R(1-\nu)}\frac{\partial M_T}{\partial y} \\ \overline{Z} &= Z - \frac{N_T}{R(1-\nu)} - \frac{1}{1-\nu}\nabla^2 M_T \end{aligned}\right\} \quad (6.30)$$

Accordingly, the following very simplified set of displacement equations may be written,[37,164]

$$\left.\begin{aligned} \frac{\partial^2 u}{\partial x^2} + \frac{1-\nu}{2}\frac{\partial^2 u}{\partial y^2} + \frac{1+\nu}{2}\frac{\partial^2 v}{\partial x\partial y} - \frac{\nu}{R}\frac{\partial w}{\partial x} &= -\overline{X}\frac{(1-\nu^2)}{Ed} \\ \frac{1+\nu}{2}\frac{\partial^2 u}{\partial x\partial y} + \frac{1-\nu}{2}\frac{\partial^2 v}{\partial x^2} + \frac{\partial^2 v}{\partial y^2} - \frac{1}{R}\frac{\partial w}{\partial y} &= -\overline{Y}\frac{(1-\nu^2)}{Ed} \\ \frac{\nu}{R}\frac{\partial u}{\partial x} + \frac{1}{R}\frac{\partial v}{\partial y} - \frac{w}{R^2} - \frac{d^2}{12R^2}\,.\,\nabla^4 w\,.\,R^2 &= -\overline{Z}\frac{(1-\nu^2)}{Ed} \end{aligned}\right\} \quad (6.31)$$

Thus, in this set of equations the operators A_{ww}, B_{ww} and C_{ww} are given by

$$\left.\begin{aligned} A_{ww} &= -\left(\frac{1}{R^2} + \frac{d^2}{12}\nabla^4\right) \\ \text{hence} \qquad B_{ww} &= -\frac{1}{R^2}\,;\; C_{ww} = -R^2\nabla^4 \end{aligned}\right\} \tag{6.32}$$

These equations (6.31) are so simple that it is quite easy to eliminate u and v from them and to arrive at a single differential equation in w, viz.

$$\nabla^8 w + \frac{12(1-\nu^2)}{R^2 d^2}\frac{\partial^4 w}{\partial x^4} = \frac{1}{D}\left[\nabla^4 \bar{Z} + \frac{1}{R}\left\{\bar{X}^{100} - \nu \bar{X}^{111} - \bar{Y}^{000} - (2+\nu)\,\bar{Y}^{110}\right\}\right] \tag{6.33}$$

$$\left.\begin{aligned} \text{where } \nabla^4 v &= \frac{(2+\nu)}{R}\frac{\partial^3 w}{\partial x^2 \partial y} + \frac{1}{R}\frac{\partial^3 w}{\partial y^3} \\ &\quad + \frac{(1+\nu)^2}{Ed}\left[\bar{X}^{10} - \frac{2}{1+\nu}\bar{Y}^{11} - \left(\frac{1-\nu}{1+\nu}\right)\bar{Y}^{00}\right] \\ \nabla^4 u &= \frac{\nu}{R}\frac{\partial^3 w}{\partial x^3} - \frac{1}{R}\frac{\partial^3 w}{\partial x \partial y^2} \\ &\quad + \frac{(1+\nu)^2}{Ed}\left[\bar{Y}^{10} - \frac{2}{1+\nu}\bar{X}^{00} - \left(\frac{1-\nu}{1+\nu}\right)\bar{X}^{11}\right] \end{aligned}\right\} \tag{6.34}$$

and the symbols $(\ )^1$, $(\ )^0$ denote differentiation with respect to x and y respectively.

If the tangential external loading is neglected ($X = Y = 0$), eqn. (6.33) becomes

$$\begin{aligned} D\nabla^8 w &+ \frac{Ed}{R^2}\frac{\partial^4 w}{\partial x^4} + \nabla^6 M_T/(1-\nu) \\ &+ \left(\frac{1}{1-\nu}\right)\frac{1}{R^2}\left[M_T^{0000} + (2+\nu)\,M_T^{1100}\right] \\ &+ \frac{1}{R}\left(\frac{\partial^4}{\partial x^4} + \frac{\partial^4}{\partial x^2 \partial y^2}\right) N_T = \nabla^4 Z \end{aligned} \tag{6.35}$$

Bijlaard[11] has presented similar results to those above, using, as a basis, a set of displacement equations more complex than (6.31).

It should also be noted that the analyses of the preceding sections also give similar results when the values $R_1 = \infty$ and $R_2 =$ constant (R) are substituted. Thus eqns. (6.4) and (6.6) may be written

$$\left.\begin{aligned} &D\nabla^4 w + \nabla^2 M_T/(1-\nu) - \frac{1}{R}\frac{\partial^2 \phi_T}{\partial x^2} = Z \\ &\text{and} \\ &\nabla^4 \phi_T + \frac{Ed}{R}\frac{\partial^2 w}{\partial x^2} + \nabla^2 N_T = 0 \end{aligned}\right\} \quad (6.36)$$

and eqn. (6.7) becomes

$$D\nabla^8 w + \frac{Ed}{R^2}\frac{\partial^4 w}{\partial x^4} + \nabla^6 M_T/(1-\nu) + \frac{1}{R}\left(\frac{\partial^4}{\partial x^4} + \frac{\partial^4}{\partial x^2 \partial y^2}\right) N_T = \nabla^4 Z \quad (6.37)$$

For cylinders of small thickness to radius ratio the term

$$\left(\frac{1}{1-\nu}\right)\frac{1}{R^2}\left[M_T^{0000} + (2+\nu)\, M_T^{1100}\right]$$

is small and eqns. (6.35) and (6.37) are then equal. It is also seen that eqn. (6.36) correspond to eqns. (6.8) and (6.9) when the non-linear terms are omitted from the latter.

A large-deflexion analysis in reference 157 produced the following set of equations;

$$D\left[\nabla^4 w + \frac{2}{R^2}\left(\frac{\partial^2 w}{\partial y^2} + \nu\frac{\partial^2 w}{\partial x^2}\right) + \frac{w}{R^4}\right] + \frac{1}{1-\nu}\left[\nabla^2 M_T + \frac{M_T}{R^2}\right]$$
$$- \frac{1}{R}\frac{\partial^2 \phi_T}{\partial x^2} - \left[\underline{\frac{\partial^2 w}{\partial x^2}\frac{\partial^2 \phi_T}{\partial y^2}} - 2\,\underline{\frac{\partial^2 w}{\partial x \partial y}\frac{\partial^2 \phi_T}{\partial x \partial y}} + \underline{\frac{\partial^2 w}{\partial y^2}\frac{\partial^2 \phi_T}{\partial x^2}}\right] = Z \quad (6.38)$$

and

$$\nabla^4 \phi_T + Ed\left[\frac{1}{R}\frac{\partial^2 w}{\partial x^2} + \underline{\frac{\partial^2 w}{\partial x^2}\frac{\partial^2 w}{\partial y^2}} - \underline{\left(\frac{\partial^2 w}{\partial x \partial y}\right)^2}\right] + \nabla^2 N_T = 0$$

where the non-linear terms are underlined. It is seen that the form of these equations is slightly different from (6.8) and (6.9). This must be due to differences in the assumed strain–displacement relations.

To satisfy the hypothesis of plane stress the same assumption is made as for flat plates, viz. that the temperature distribution $T_{(x,y,z)}$ can be expressed by (cf. (4.22))

$$T_{(x,y,z)} = T_0 + \Delta T\left(\frac{z}{d}\right) \qquad (6.39)$$

where T_0 and ΔT are functions of x, y.

To solve the above equations it is desired to find the functions u, v and w satisfying the equations (6.31), for the particular temperature and external load distributions. For simply supported shells this is quite straightforward since X, Y, Z, T_0, ΔT, u, v, and w are all expressed as double Fourier series and the coefficients in the displacement series are then evaluated by comparing corresponding terms. For other boundary conditions the procedure is not so obvious, but, by analogy between the thermal and mechanical loading terms in $\overline{X}$, $\overline{Y}$ and $\overline{Z}$, procedures for isothermal analyses may be invoked.

It should be remembered that all solutions obtained directly from the eighth-order equations above may *not* be acceptable. Since these equations have been derived by differentiation of the original displacement equations, e.g. (6.35) obtained from (6.31), the only solutions acceptable are those which also satisfy the original equations, e.g. (6.31).

Because eqns. (6.31) are very much simplified it follows that the conditions under which it may be intended to use (6.31) should be examined carefully to ensure that they do not invalidate any of the assumptions made in deriving (6.31). Equations (6.31) and (6.35) are equivalent to Donnell's simplified equations and a well-known limitation on their use applies when the number of circumferential waves (n) in the deformation mode for w is in the range $1 < n \leqslant 3$.

The above analyses can be used for circular cylindrical shells subjected to arbitrary temperature distributions of the form given in (6.39). In the following sections some of the effects of radial, circumferential and axial temperature distributions will be considered separately. The subject of discontinuity stresses is also considered in this Chapter; these stresses arise near the junction of the shell with any other member having different thermal expansions.

6.5.2. Radial and Circumferential Temperature Distributions

Goodier[54,56] presents an analysis for this combined problem assuming that the radial temperature gradient is linear and the circumferential variation can be expressed by a Fourier series. Thus if T_1 and T_2 are the temperatures on the inner and outer surfaces of the shell, where

$$\left.\begin{aligned} T_1 &= A_0 + A_1 \cos\theta + \ldots + B_1 \sin\theta + \ldots \\ T_2 &= A_0' + A_1' \cos\theta + \ldots + B_1' \sin\theta + \ldots \end{aligned}\right\} \tag{6.40}$$

and θ is the angular coordinate, the thermal stresses can be written in terms of the above coefficients only, without the necessity of considering any further terms in the series.[15]

Since the temperature distribution is not a function of the axial coordinate a plain strain analysis is possible and, accordingly, the stresses σ_{rr}, $\sigma_{r\theta}$, $\sigma_{\theta\theta}$ and σ_{xx} may be determined. Goodier's analysis has a different basis but, apart from an error which has been pointed out by Tsao[166], the results obtained are of sufficient accuracy for thin shells. Thus, the extreme circumferential stress, $\sigma_{\theta\theta}$, is given by,

$$\sigma_{\theta\theta} = \pm \frac{Ea}{2(1-\nu)}\left[(A_0 - A_0') + (A_1 - A_1')\cos\theta + (B_1 - B_1')\sin\theta\right] \tag{6.41}$$

for which the maximum value is one of the two quantities,

$$(\sigma_{\theta\theta})_{\max} = \frac{Ea}{2(1-\nu)}\left[(A_0 - A_0') \pm \{(A_1 - A_1')^2 + (B_1 - B_1')^2\}^{\frac{1}{2}}\right] \tag{6.42}$$

depending upon whether A_0 is greater or less than A_0'. The corresponding expression for the axial stress at a distance from the free ends of the shell is[54,56,166]

$$\sigma_{xx} = Ea\left[-T_m \pm \tfrac{1}{2}(T_2 - T_1) + \tfrac{1}{2}(A_0 + A_0')\right] \pm \frac{Ea\nu}{2(1-\nu)}[\Lambda] \tag{6.43}$$

where [Λ] is the expression given in (6.41). In eqns. (6.41)–(6.43) the positive sign refers to the outer surface of the shell.

It is seen that all these expressions for thin shells are independent of the shell size or thickness. This result was noted in Section 5.7 and it can be shown that the results of the present section agree with those of section 5.7 for radial temperature distributions only, viz.

$$(\sigma_{\theta\theta})_{max} = (\sigma_{xx})_{max} = \pm\, E\alpha\,(A_0 - A_0')/2(1 - \nu) \qquad (6.44)$$

Near the free ends of the shell this uniform stress state is disturbed by the requirement that the stress normal to the free edge (σ_{xx}) be zero.[164] Because of this condition the local value of the circumferential stress at the free ends is increased to a value approximately 25 per cent greater than that in eqn. (6.44).

If it is assumed that the shell is heated uniformly over a small arc of circumferential length λ along the length of its outer surface the temperature distribution may be approximated by T_2, constant over λ, with the rest of the outer surface and the whole inner surface kept at T_1 (constant). This example is analyzed by Goodier[54,56] and corrected by Tsao[166]. It may be shown that the maximum circumferential stress is given by

$$(\sigma_{\theta\theta})_{max} = \frac{3}{4}\,\frac{E\alpha(T_2 - T_1)}{(1 - \nu)}\cdot\frac{\lambda}{R} \qquad (6.45)$$

but for such localized heating λ/R is small and large circumferential stresses are not produced. The corresponding expression for the axial stress, away from the free ends, is[166]

$$\sigma_{xx} = \frac{-\,E\alpha(T_2 - T_1)}{2}\left[1 - 2\left(\frac{z}{d}\right) - \frac{3\lambda\nu}{R(1 - \nu)}\left(\frac{z}{d}\right)\right] \qquad (6.46)$$

where the shell lies between $z = d/2$ (inside) and $-\,d/2$ (outside). For localized heating λ/R is small and eqn. (6.46) reduces to the simple result that the axial stress varies linearly through the thickness from zero at the inside to $E\alpha\,(T_2 - T_1)$ at the outside (compression).

6.5.3. Axial Temperature Distributions

The effects of arbitrary, axial, temperature distributions are considered in some detail by Kent[91] who assumes the temperature distribution to be of general polynomial form. Den Hartog[23] gives an analysis to the problem of an infinitely long shell with a sinusoidal temperature distribution along its length, and he also considers the finite length shell with either a single half-sine-wave distribution, or a continuous zigzag distribution.

For a temperature gradient and pressure variation in the axial direction only, the problem is axisymmetric and for small deflexions, eqns. (6.36) and (6.38) produce respectively,

$$D\left[\frac{d^4w}{dx^4}\right] = Z_{(x)} + \frac{1}{R}\frac{d^2\phi_T}{dx^2} \tag{6.47}$$

$$D\left[\frac{d^4w}{dx^4} + \frac{2\nu}{R^2}\frac{d^2w}{dx^2} + \frac{w}{R^4}\right] = Z_{(x)} + \frac{1}{R}\frac{d^2\phi_T}{dx^2} \tag{6.48}$$

where ϕ_T is given in both cases by

$$\frac{d^4\phi_T}{dx^4} = -\frac{d^2N_T}{dx^2} - \frac{Ed}{R}\frac{d^2w}{dx^2} \tag{6.49}$$

and $N_T = E\alpha T_{(x)}d$. It can be shown by an order of magnitude analysis that the additional terms in (6.48) compared with (6.47) are negligible for typical shell d/R ratios of practical interest. Thus, by solving for ϕ_T from (6.49) and substituting into either (6.47) or (6.48) the following equation is obtained

$$\frac{d^4w}{dx^4} + 4\beta^4 w = \frac{Z_{(x)}}{D} - \frac{E\alpha d}{DR}T_{(x)} \tag{6.50}$$

where $4\beta^4 = 12(1-\nu^2)/d^2R^2$.

Therefore the temperature term may be considered as equivalent to a transverse pressure normal to the shell surface.[164] The general solution of this equation may be written as either

$$w = e^{\beta x}[C_1\cos\beta x + C_2\sin\beta x] + e^{-\beta x}[C_3\cos\beta x + C_4\sin\beta x] + P_{(x)} \tag{6.51}$$

or

$$w = K_1\sin\beta x\sinh\beta x + K_2\sin\beta x\cosh\beta x + K_3\cos\beta x\sinh\beta x + K_4\cos\beta x\cosh\beta x + P_{(x)} \tag{6.52}$$

in which $P_{(x)}$ is a particular solution of (6.50) which depends on the form of $Z_{(x)}$ and $T_{(x)}$, and the constants $C_1 \ldots C_4$ or $K_1 \ldots K_4$ depend on the boundary conditions at the ends of the shell.

For known boundary conditions the above equations may be solved directly for the thermoelastic problem, e.g. for a free edge the appropriate conditions to be satisfied are zero edge moment (M_x) and shear force (Q_x), i.e. $\mathrm{d}^2w/\mathrm{d}x^2 = \mathrm{d}^3w/\mathrm{d}x^3 = 0$[164]; similarly for a fully clamped edge $w = \mathrm{d}w/\mathrm{d}x = 0$. Indeterminate boundary conditions occur when the shell is attached to a flexible frame or bulkhead and the procedure then is to satisfy conditions of compatibility of load and displacement at the attachment points by using influence coefficients for the adjacent structural members. This latter problem is classed as a *discontinuity problem* and is considered in more detail in the next section.

For known boundary conditions typical solutions to (6.50) are now given,[88] with the assumption that $Z = 0$.

I. A long shell has a free edge at one end ($x = 0$) and is subjected to a polynomial temperature distribution of the form

$$T_{(x)} = \sum_{m=0}^{3} F_m x^m.$$

In this case the particular solution of (6.50) is given by $w_{(x)} = -Ra T_{(x)}$ and the general solution of (6.50) is found using (6.51). For long shells this form of solution is more suitable than (6.52) since, obviously, $C_1 = C_2 = 0$ if the deformations of the shell are to be finite at large values of x. Thus the required form of solution is

$$w = e^{-\beta x}\left[C_3 \cos \beta x + C_4 \sin \beta x\right] - Ra T_{(x)} \tag{6.53}$$

which for free edge boundary conditions becomes

$$w = \frac{aR}{2}\left\{e^{-\beta x}\left[\cos \beta x\left(\frac{T_0''}{\beta^2} + \frac{T_0'''}{\beta^3}\right) - \sin \beta x \frac{T_0''}{\beta^2}\right] - 2\, T_{(x)}\right\} \tag{6.54}$$

At $x = 0$ the values of the free edge displacement and rotation due to the temperature distribution $T_{(x)}$, are

$$\left.\begin{aligned} w_0{}^T &= \frac{aR}{2}\left[\frac{T_0''}{\beta^2}+\frac{T_0'''}{\beta^3}-2T_0\right] \\ \theta_0{}^T &= -\frac{aR\beta}{2}\left[\frac{2T_0''}{\beta^2}+\frac{T_0'''}{\beta^3}+\frac{2T_0'}{\beta}\right] \end{aligned}\right\} \tag{6.55}$$

In the expressions above T_0''' is defined as the value of $\mathrm{d}^3T_{(x)}/\mathrm{d}x^3$ at $x = 0$, and the other terms T_0, T_0' and T_0'' are similarly defined. Since the hoop stress σ_{yy} is given by $(1/d)(\mathrm{d}^2\phi_T/\mathrm{d}x^2)$ it follows from (6.47) and (6.49) that

$$\sigma_{yy} = \frac{DR}{d}\frac{\mathrm{d}^4w}{\mathrm{d}x^4} = -\frac{E}{R}(w + aRT_{(x)}) \tag{6.56}$$

II. A long shell is clamped at $x = 0$ and is subjected to a uniform temperature rise T. The edge conditions are $w = \mathrm{d}w/\mathrm{d}x = 0$ at $x = 0$.

In this case the solution to (6.51) is

$$w = -aRT[1 - e^{-\beta x}(\cos\beta x + \sin\beta x)] \tag{6.57}$$

from which the following expression for σ_{yy} is obtained;

$$\sigma_{yy} = -Ea T\mathrm{e}^{-\beta x}(\cos\beta x + \sin\beta x) \tag{6.58}$$

III. A short shell has free edges at $x = \pm L/2$ and is subjected to a symmetrical temperature distribution about the mid-length of the shell, $x = 0$, with the form

$$T_{(x)} = \sum_{m=0}^{3} F_m|x|^m.$$

In this case the general solution is taken in the form of (6.52), where by symmetry $K_2 = K_3 = 0$, and K_1 and K_4 are found to be

$$\begin{aligned} K_1 &= aR\left[\frac{T_E''}{\beta^2}(\sin\gamma\cosh\gamma + \cos\gamma\sinh\gamma) - \frac{T_E'''}{\beta^3}\sin\gamma\sinh\gamma\right]\Big/\left[\sinh 2\gamma + \sin 2\gamma\right] \\ K_4 &= aR\left[\frac{T_E''}{\beta^2}(\cos\gamma\sinh\gamma - \sin\gamma\cosh\gamma) - \frac{T_E'''}{\beta^3}\cos\gamma\cosh\gamma\right]\Big/\left[\sinh 2\gamma + \sin 2\gamma\right] \end{aligned} \tag{6.59}$$

The corresponding values of edge displacement and rotation are given by

$$\left.\begin{aligned} w_E^T &= K_1 \sin\gamma \sinh\gamma + K_4 \cos\gamma \cosh\gamma - \alpha R T_E \\ \theta_E^T &= \pm\,\beta \left\{ K_1 [\sin\gamma \cosh\gamma + \cos\gamma \sinh\gamma] \right. \\ &\left. + K_4 [\cos\gamma \sinh\gamma - \sin\gamma \cosh\gamma] - \frac{\alpha R T_E'}{\beta} \right\} \end{aligned}\right\} \quad (6.60)$$

In the expressions above T_E'', for example, is defined as the value of $\mathrm{d}^2 T_{(x)}/\mathrm{d}x^2$ at the end of the shell where $x = \pm\, L/2$; also $\gamma = \beta L/2$.

The above examples illustrate typical procedures for calculating the thermoelastic radial deformations of a circular cylindrical shell with known boundary conditions. Other similar analyses for polynomial and trigonometric temperature distributions are found in references 74, 81 and 84.

6.6. Discontinuity Problems in Shells

The various structural elements considered in this text, e.g. plates, discs, beams or shells, are generally components of a complete structure and, therefore, compatibility of load and displacement must be satisfied at the joints between various elements. When such elements offer flexible support one to the other the satisfaction of compatibility constitutes what is often termed a *discontinuity problem*.

To illustrate the analytical procedure a discontinuity problem is analyzed for two, dissimilar, circular cylindrical shells, with axial temperature distributions, rigidly clamped to a cooler (or hotter) flexible plane bulkhead. This configuration is commonplace in aircraft and nuclear engineering structures (amongst others) and several published papers present analyses on this topic.

Przemieniecki[140] considers a shell of infinite length stiffened by a plane, constant-thickness bulkhead with a uniform temperature rise in the shell and a radial variation of temperature in the bulkhead. In references 81, 87 and 88, this analysis is extended to allow a polynomial temperature distribution along the shell length (see Section 6.5.3) and in another paper[84] some effects of finite shell length are examined. The threefold discontinuity problem near the junction of two dissimilar circular shells and a plane bulkhead (or frame) due to pressure and/or thermal effects, is considered in reference 85, and the corresponding problem at the junction of a circular cylindrical shell and a sphere is analyzed in reference 73. An approach similar to reference 85 is outlined in reference 159 which also considers the same problem for a conical shell.

The configuration to be considered is shown in Fig. 6.5 and it is assumed initially that the various members are unconnected. The deflexions and slopes at the edges of the unattached shells and bulkhead are first determined for the thermal " loading " only, using, for example, the analyses of Sections 6.5.3, 3.2, and 4.5.1. Thus,

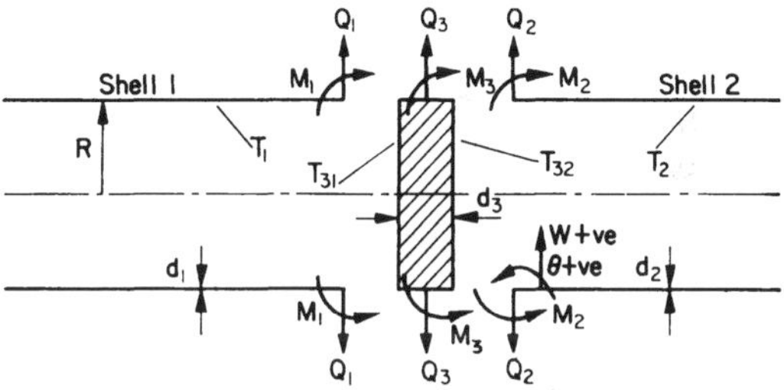

FIG. 6.5. Bulkhead-stiffened circular cylindrical shell.

eqn. (6.55) for long shells enables expressions to be determined for w_{01}^T, w_{02}^T, θ_{01}^T, θ_{02}^T at the shell edges. For the bulkhead an axisymmetric temperature distribution may be assumed giving a temperature difference across the bulkhead thickness $\Delta T \; (= F_{(r)})$ and a mean radial temperature distribution $T_3 (= f_{(r)})$. If T_{31} and T_{32} are the arbitrary radial temperature distributions on opposite faces of the bulkhead with linear variation across the thickness then

$$\left.\begin{aligned} \Delta T &= T_{31} - T_{32} = F_{(r)} \\ T_3 &= \tfrac{1}{2}(T_{31} + T_{32}) = f_{(r)} \end{aligned}\right\} \tag{6.61}$$

From Section 4.5.1, eqn. (4.63), it follows that the radial extension and edge rotation due to ΔT are,

$$w_3^T = 0 \; ; \; \theta_3^T = (2a_3/Rd_3) \int_0^R \Delta T r \, dr \tag{6.62}$$

and from Section 3.2, eqn. (3.22), the expressions corresponding to T_3 are

$$w_3^T = -(2a_3/R) \int_0^R T_3 r \, dr \; ; \; \theta_3^T = 0 \tag{6.63}$$

The combined expressions for w_3^T and θ_3^T are easily determined.

Because the various expressions w^T_{01}, w^T_{02}, w^T_3 and θ^T_{01}, θ^T_{02}, θ^T_3 are not necessarily compatible, the self-equilibrating system of edge moments and shears in Fig. 6.5 is established to ensure compatibility which requires that for a clamped shell-bulkhead joint,

$$w_{01} + w^T_{01} = w_{02} + w^T_{02} = w_3 + w^T_3 \tag{6.64}$$

$$\theta_{01} + \theta^T_{01} = \theta_{02} + \theta^T_{02} = \theta_3 + \theta^T_3 \tag{6.65}$$

and

$$Q_1 + Q_2 + Q_3 = 0 \tag{6.66}$$

$$M_1 + M_2 + M_3 = 0 \tag{6.67}$$

(For simply supported edge conditions only eqns. (6.64) and (6.66) need to be satisfied.) The relationships between the edge displacements and rotations (w_{01}, θ_{01}, etc.) and the corresponding edge shears and moments (Q_1, M_1, etc.) are determined by isothermal shell and plate theory,[164] and, using previously defined notation, may be written as

$$\left.\begin{aligned} w_{01} &= -[1/2\beta_1^3 D_1][-\beta_1 M_1 + Q_1] \\ \theta_{01} &= -[1/2\beta_1^2 D_1][-2\beta_1 M_1 + Q_1] \end{aligned}\right\} \tag{6.68}$$

$$\left.\begin{aligned} w_{02} &= -[1/2\beta_2^3 D_2][\beta_2 M_2 + Q_2] \\ \theta_{02} &= [1/2\beta_2^2 D_2][2\beta_2 M_2 + Q_2] \end{aligned}\right\} \tag{6.69}$$

and

$$\left.\begin{aligned} w_3 &= -Q_3(1 - \nu_3)R/E_3 d_3 \\ \theta_3 &= M_3 R/D_3(1 + \nu_3) \end{aligned}\right\} \tag{6.70}$$

Equations (6.68) and (6.69) only apply to long shells and may be obtained from eqn. (6.51) with $P_{(x)} = 0$ and $C_1 = C_2 = 0$.[164] Similar expressions for short shells are given in references 88 and 164.

Substitution of the expressions for w^T_{01}, w_{01}, θ^T_{01}, θ_{01} etc. into eqns. (6.64) and (6.65) and the use of (6.66), (6.67) enable the values of the six unknown edge moments and shears to be determined. The stress distribution corresponding to these edge moments and shears may be found from isothermal analyses and added to those previously determined for the unattached shells and bulkhead.

Thus, the discontinuity problem is solved for shells which are rigidly clamped to a plane bulkhead. For simply supported attachments the necessary modifications to the analysis above are obvious. Extensions of the above procedure to other classes of problems and types of structure can be readily visualized and various examples may be found in the references quoted earlier.

CHAPTER 7

Thermal Buckling

7.1. Introduction

It is observed from the analyses of previous chapters that thermal stresses in structures are self-equilibrating, with the result that both compressive and tensile thermal stresses occur. If the structure is slender the possibility of thermal buckling due to the compressive thermal stresses must be examined. Since many compression elements fail at stresses in the inelastic region, even at room temperatures, it follows that a moderate to high but uniform rise in temperature only affects the problem through changes in the elastic limit and modulus of the material. In fact a complete knowledge of the material stress–strain curve is required and since this may be a function of temperature for the material concerned (and also of time at temperature, rate of rise of temperature) a large amount of experimental data is required before realistic analyses can be undertaken. This point has been emphasized in papers by Hoff(70) and Van der Neut(169).

The existing stability theories for various structural elements were originally derived for buckling under external loads of known magnitude. The magnitude of thermal stresses however, is greatly influenced by the relative internal stiffnesses, restraints and deformations of the structure, so it is not immediately obvious that the accepted methods of stability analysis can be applied directly to thermal buckling problems. However, it is reported in reference 70 that an analysis of a strut, which is heated uniformly whilst its ends are maintained at a fixed distance apart, shows that first-order small lateral displacements of the strut cause only second-order small changes in the thermal stresses at the moment of buckling. In this case the classical method of analysis based on prescribed loads gives satisfactory results and it may be assumed that thermal buckling problems of plates and shells may also be analyzed using methods developed for isothermal problems.

It should be remembered that a structure may be capable of sustaining load (temperatures) greater than the buckling load (temperature) although the effective stiffness of the structure is then lower after buckling. Whilst this may be acceptable from a strength point of view, the lateral deflexions caused may not be permissible for other reasons, e.g. waviness of a wing surface may impair its aerodynamic performance. A more complete discussion of such problems and their effects on aircraft structures is given in reference 169.

Even if the thermal stresses alone, or the stresses due to mechanical loads alone, do not cause buckling, the combination of both may prove critical. Since both types of stress are different functions of the structural configuration this leads to problems of optimum design to sustain the combined loading (see Chapter 8). It must be emphasized that thermal buckling is not immediately catastrophic since the greater flexibility after buckling inhibits further increase in thermal stress level.

7.2. Columns or Bars

If an initially straight uniform bar is subjected to a uniform temperature rise T, and it is supported in such a way that its length cannot change, the compressive thermal stress produced is given by $E\alpha T$. When this value reaches the Euler critical stress for the bar, buckling begins. The lateral post-buckling deflexions can then be determined by assuming that the mean compressive strain remains constant and that any further increase in the thermal expansion of the bar produces only an increase in the lateral deformation.

If the bar of length L is simply supported at its ends the possibility of Euler buckling as a strut may be examined by means of the well known formula

$$\sigma_{cr} = E(\alpha T)_{cr} = \pi^2 EI/AL^2 \tag{7.1}$$

where A is the cross-sectional area of the bar and I is its least Second Moment of Area. Thus

$$(\alpha T)_{cr} = \pi^2/(L/\rho)^2 \tag{7.2}$$

where ρ is the least radius of gyration of the cross-section. It is seen that the critical thermal strain $(\alpha T)_{cr}$ is independent of E, the Elastic Modulus of the material. This deduction from the very simple example considered may be accepted as generally true for all thermal buckling problems, apart from the effects of variations in modulus due to temperature or thermal stress level. Thus the intuitively obvious result is obtained that a material with a lower value of α is better from the thermal buckling point of view.

For a simply supported column or bar which is axially restrained a very concise analysis is given by Gatewood[48]. The initial eccentricity in the bar (W_i) and the total lateral deflexion (W_T) are given by

$$\left.\begin{aligned} W_i &= W_{mi} \sin (\pi x/L) \\ W_T &= W_{mT} \sin (\pi x/L) \end{aligned}\right\} \tag{7.3}$$

which are related approximately by the expression[162]

$$W_{mT} = W_{mi}/(1 - e_0/e_{cr}) \tag{7.4}$$

where e_0 is the mean compressive strain and $e_{cr} = (\alpha T)_{cr}$. Since there is a change in length of the bar due to lateral deflexions, e_0 may be written as

$$e_0 = \alpha T - \Delta L/L = \alpha T - \left(\frac{\pi W_{mT}}{2L}\right)^2 + \left(\frac{\pi W_{mi}}{2L}\right)^2 \tag{7.5}$$

which may be rearranged to give

$$R_T = \frac{\alpha T}{(\alpha T)_{cr}} = \frac{e_0}{e_{cr}} + D_T^2 - D_i^2 \tag{7.6}$$

where $\quad D_T = \pi W_{mT}/2L(e_{cr})^{1/2}$ and $D_i = \pi W_{mi}/2L(e_{cr})^{1/2}$.

When $W_{mi} = 0$ and R_T is less than unity then from (7.4) and (7.6), $W_{mT} = 0$ and e_0/e_{cr} is linear with R_T. In this case a value of R_T greater than unity produces the result that e_0/e_{cr} stays constant at unity and

$$D_T = (R_T - 1)^{1/2} \tag{7.7}$$

Therefore, although the temperature T_{cr} is called the buckling temperature it does not represent a structural failure, it merely indicates that the mean compressive strain has reached a maximum value, $(\alpha T)_{cr}$, and that lateral deflexions, W_T are initiated. When $W_{mi} \neq 0$ lateral deflexions are increasing continuously with T from $T = 0$ so that there is not such an abrupt change in behaviour at T_{cr} as for $W_i = 0$.

A consequence of the lateral deflexions is the introduction of bending stresses along the bar and the possibility of yielding in the compression fibres of the bar. The maximum (yield) strain in this case is given by

$$e_y = e_0\left[1 + A\,(W_{mT} - W_{mi})\left(\frac{c}{I}\right)\right] \tag{7.8}$$

from which

$$D_{Ty} = D_i + \frac{\rho}{2c}\left(\frac{e_y}{e_{cr}} \cdot \frac{e_{cr}}{e_0} - 1\right) \tag{7.9}$$

where c is the distance of the farthest compression fibre from the neutral axis. This equation enables the temperature to be predicted at which yielding occurs, for a given strut configuration $(\rho/2c)$ and initial eccentricity D_i $(= W_{mi}/2\rho)$, when it is used in conjunction with the appropriate forms of (7.6) and (7.4), viz.

$$R_{Ty} = \frac{e_0}{e_{cr}} + D_{Ty}^2 - D_i^2 \tag{7.10}$$

$$\frac{e_0}{e_{cr}} = 1 - \frac{D_i}{D_{Ty}} \tag{7.11}$$

In the general case substitution of (7.11) into (7.9) yields a quadratic in D_{Ty} from which D_{Ty} and then R_{Ty} (in (7.10)) are determined. Thus if $e_y/e_{cr} = 4{\cdot}0$, $D_i = 0{\cdot}1$ and $\rho/2c = \frac{1}{3}$, then $D_{Ty} = 1{\cdot}426$, $e_0/e_{cr} = 0{\cdot}93$ and $R_{Ty} = 2{\cdot}95$.

The above analyses are for a uniform temperature rise in the bar. If the bar has temperature variations over the cross-section or along the length the analysis is more cumbersome, and the reader should consult references 15 or 158 for a more general derivation.

7.3. Thermal Buckling of Flat, Uniform Plates

7.3.1 Introduction

The governing equation of equilibrium for a uniform plate loaded only in its plane is, from (4.25),

$$D\nabla^4 w = d\left[\sigma_{xx}\frac{\partial^2 w}{\partial x^2} + 2\sigma_{xy}\frac{\partial^2 w}{\partial x \partial y} + \sigma_{yy}\frac{\partial^2 w}{\partial y^2}\right] \tag{7.12}$$

where the membrane stresses σ_{xx} etc. are determined from the compatibility equation (4.15), i.e. $\sigma_{xx} = \partial^2\phi/\partial y^2$ etc., where

$$\nabla^4\phi = -\frac{1}{d}\nabla^2 N_T \tag{7.13}$$

These equations are valid for an initially flat plate subjected only to small deflexions and they show that the function ϕ, and hence the stress distribution, is not affected by the displacement w at the onset of buckling.

Since, in general, the thermal stress distributions are not uniform the exact solution of (7.12) is difficult to obtain. This is particularly true for plates of arbitrary shape and when due allowance is made for the exact boundary conditions. For these general situations the methods of Ritz–Galerkin or Rayleigh–Ritz are recommended.[169] The latter method is based on the principle of minimum potential energy, but to avoid the necessity of actually calculating the terms in the energy expression, the former method which is physically identical appears preferable; it is simpler to apply and is derived directly from the governing equilibrium equation (7.12). Both methods will now be presented.

7.3.2. The Ritz–Galerkin Method

The deflexion w is assumed to have the form

$$w = \sum_{i=1}^{N} C_i w_{i(x\,y)} \tag{7.14}$$

where the terms C_i are, initially, undetermined coefficients for the displacement functions $w_{i(x,y)}$. These functions must satisfy the kinematic conditions on the plate boundaries but they need not necessarily satisfy the dynamic boundary conditions, i.e. the functions for a clamped edge **must** satisfy the conditions of zero displacement and zero slope but agreement with the values of edge shear and bending moment, whilst it is desirable, is not essential.

On substituting (7.14) into (7.12) it is found, in general, that the latter equation is not satisfied exactly at all points of the plate. The error functions involved may be visualized as a " fictitious pressure " q which is given by

$$q = D\nabla^4 w - d\left[\sigma_{xx}\frac{\partial^2 w}{\partial x^2} + 2\sigma_{xy}\frac{\partial^2 w}{\partial x\,\partial y} + \sigma_{yy}\frac{\partial^2 w}{\partial y^2}\right] \tag{7.15}$$

together with " additional loads " at the boundary which must be considered as acting when (7.14) does not satisfy the dynamic boundary conditions, viz.

$$\left.\begin{aligned} M_{nn} &= -D\left[\frac{\partial^2 w}{\partial n^2} + \nu\frac{\partial^2 w}{\partial s^2}\right] \\ Q &= -D\left[\frac{\partial^3 w}{\partial n^3} + (2-\nu)\frac{\partial^3 w}{\partial n \partial s^2}\right] + d\left[\sigma_{nn}\frac{\partial w}{\partial n} + \sigma_{ns}\frac{\partial w}{\partial s}\right] \end{aligned}\right\} \tag{7.16}$$

In order to minimize the potential energy the work done by q, M_{nn} and Q, when acting through virtual displacements w_i, must be equated to zero. Thus,

$$\iint q\, w_i \mathrm{d}x\mathrm{d}y - \int M_{nn}\left(\frac{\partial w_i}{\partial n}\right)\mathrm{d}s + \int Q \cdot w_i \mathrm{d}s = 0 \;(i = 1, 2, \ldots N) \tag{7.17}$$

and the buckling condition is found when the determinant Δ formed by the N equations of (7.17) is zero. If the stresses σ_{xx}, σ_{xy} and σ_{yy} are expressed in terms of critical coefficients C_{1m}, C_{2m} and C_{3m} respectively, the roots of the equation $\Delta = 0$ yield the critical values of C_m. Since the w_i may be expressed in terms of arbitrary buckle wavelength parameters, m, it follows that the C's must be examined carefully to ensure that the values of the parameters m yield the most critical values of σ_{xx}, σ_{xy} and σ_{yy}.

An application of this procedure to a simply supported rectangular plate is illustrated in detail and by an example in reference 169.

7.3.3. The Rayleigh–Ritz Method

At the instant of buckling the total change of energy of the system is composed of the bending strain energy of the plate plus the total work done by the forces acting in the middle plane of the plate during buckling. It is assumed that the middle plane stresses do not change during buckling hence the stresses acting in supporting structure also do not change. The total change of potential energy is

$$L = \tfrac{1}{2}\iint D\left[\left(\frac{\partial^2 w}{\partial x^2} + \frac{\partial^2 w}{\partial y^2}\right)^2 - 2(1-\nu)\left\{\frac{\partial^2 w}{\partial x^2}\frac{\partial^2 w}{\partial y^2} - \left(\frac{\partial^2 w}{\partial x \partial y}\right)^2\right\}\right]\mathrm{d}x\mathrm{d}y$$
$$+ \tfrac{1}{2}\iint d\left[\sigma_{xx}\left(\frac{\partial w}{\partial x}\right)^2 + 2\sigma_{xy}\frac{\partial w}{\partial x}\frac{\partial w}{\partial y} + \sigma_{yy}\left(\frac{\partial w}{\partial y}\right)^2\right]\mathrm{d}x\mathrm{d}y \tag{7.18}$$

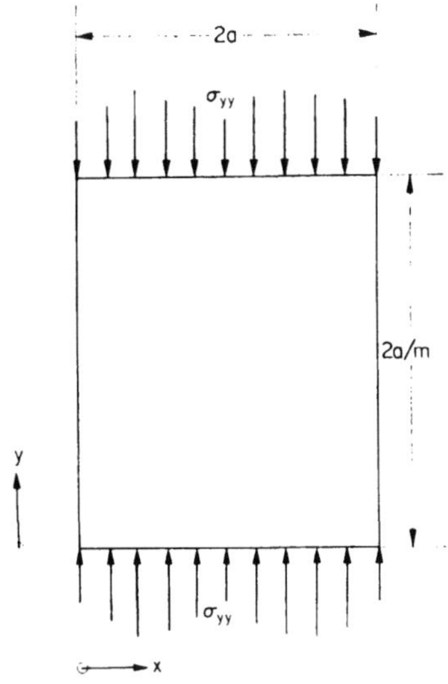

Fig. 7.1. Plate subjected to thermal stresses.

where the surface integrals are evaluated over the plate area. It should be noted that eqn. (7.18) allows of the possibility of D and d being variables.

The deflexion function w is expressed by the series (7.14) and substituted into (7.18). Since L must be stationary at buckling then $\partial L/\partial C_i = 0$ $(i = 1, 2, \ldots, n)$ which leads to N linear homogeneous simultaneous algebraic equations in the C_i. The determinant formed by the coefficients of the C_i terms vanishes at buckling. As in the Ritz–Galerkin method, the functions w_i will contain unknown buckle wavelength parameters; these must be determined so as to ensure that the minimum critical stresses for buckling are evaluated.

7.3.4. Rectangular Plates

Many analyses are available in the literature for the thermal buckling problem of thin rectangular plates, their main differences being in the form of the temperature distribution over the plate, and in the edge restraint conditions.

For a uniform plate which is simply supported along its edges and for which the only stress of importance is, say, σ_{yy}, the following procedure may be adopted (it is assumed that σ_{yy} is only a function of x).[70]

Trigonometric series are used to represent both σ_{yy} and w, viz.

$$\sigma_{yy} = \frac{-\sigma_0}{C} \sum_{r=0}^{\infty} p_r \cos(r\pi x/2a) \tag{7.19}$$

$$w = \sin\left(\frac{m\pi y}{2a}\right) \sum_{k=1}^{\infty} a_k \sin(k\pi x/2a) \tag{7.20}$$

and on substituting into (7.12) the following infinite determinant is obtained directly from the coefficients of like terms in the trigonometric series

$$\begin{vmatrix} 2(p_0 - CK_1) - p_2, & p_1 - p_3 & , & p_2 - p_4 & , \dots \\ p_1 - p_3 & , 2(p_0 - CK_2) - p_4, & p_1 - p_5 & , \dots \\ p_2 - p_4 & , \quad p_1 - p_5 & , 2(p_0 - CK_3) - p_6, & \dots \\ \cdot & \cdot & \cdot & \\ \cdot & \cdot & \cdot & \\ \cdot & \cdot & \cdot & \end{vmatrix} = 0 \tag{7.21}$$

The condition for buckling is that this determinant should vanish. In (7.19), (7.20) C is the critical buckling parameter, $2a/m$ is the half wavelength of the buckle mode in the longitudinal direction (Fig. 7.1),

$$K_s = \left[\frac{m^2 + s^2}{m}\right]^2$$

and

$$\sigma_0 = \frac{\pi^2 E}{12(1 - \nu^2)} \left(\frac{d}{2a}\right)^2 \tag{7.22}$$

Initially both C and m are unknown and eqn. (7.21) must be expanded and analyzed to find the value of m leading to the maximum value of C and, hence to the most critical buckling stress distribution (7.19). It should be noted that eqn. (7.20) satisfies the following boundary conditions

$$\left.\begin{aligned} w = \frac{\partial^2 w}{\partial x^2} = 0 \qquad & x = 0,\ 2a \\ w = \frac{\partial^2 w}{\partial y^2} = 0 \qquad & y = 0,\ 2a/m \end{aligned}\right\} \tag{7.23}$$

and eqn. (7.21) could also have been obtained by application of the Ritz–Galerkin method. For this particular example however the present simple method of analysis is available. Hoff[69,70] uses this method to investigate the effect of deviations from uniformity in the stress distribution on the buckling behaviour. He considers only the symmetric buckling problem, i.e. $p_r = 0, 2, 4, \ldots, k = 1, 3, 5 \ldots,$ which eliminates the even rows and columns from (7.21), and retains only the first two terms in each series, viz. p_0, p_2, k_1 and k_3. Equation (7.21) becomes

$$\begin{vmatrix} 2(p_0 - CK_1) - p_2, & p_2 \\ p_2 & , \quad 2(p_0 - CK_3) \end{vmatrix} = 0 \tag{7.24}$$

Assuming that $p_0 = 1$ and $p_2 = 0$ the minimum uniform buckling stress distribution is given by $C = \frac{1}{4}$ when $m = 1$; when $p_2 \neq 0$ the approximate result below is obtained for $m = 1$,

$$C = \tfrac{1}{4}\left(1 - \frac{p_2}{2}\right)(1 + \epsilon) \tag{7.25}$$

where

$$\epsilon^{-1} = \left(25 - \frac{98}{p_2} + \frac{96}{p_2^2}\right) \tag{7.26}$$

As a first approximation to (7.24) the leading diagonal term yields

$$C = \tfrac{1}{4}\left(1 - \frac{p_2}{2}\right) \tag{7.27}$$

which Hoff[70] showed is sufficiently accurate for most engineering purposes when $-2 < p_2 < 1$.

For rectangular simply supported plates a more general form for the assumed buckle mode w is

$$w = \sum_{m\,\mathrm{odd}} \sum_{n\,\mathrm{odd}} a_{mn} \sin \frac{sm\pi x}{2a} \sin \frac{tn\pi y}{2b} \tag{7.28}$$

which is symmetrical or antisymmetrical in the appropriate direction depending upon whether s and t (which are positive integers) are odd or even. In evaluating the stability determinant the problem is to determine the parameters s and t so as to minimize the stresses necessary to maintain the deformation w for the particular temperature distribution and plate aspect ratio, a/b. Since the stresses can be expressed in terms of temperature distribution parameters it is the minima of these latter quantities that are sought.

In reference 57 a temperature distribution of the form (see eqn. (3.35))

$$T_{(y)} = T_0 - \Delta T \left|\frac{y}{b}\right| \tag{7.29}$$

is chosen, i.e. a distribution of tent-like form. The rectangular plate of uniform thickness has an aspect ratio of 1·57 and the four edges of the plate are simply supported offering restraint only to lateral displacement, w. The membrane stresses are determined by the method of Section 3.6 (Method 4) and the Rayleigh–Ritz procedure is followed for the buckling analysis. The critical temperature parameter is $\Phi = \Delta T_{\mathrm{crit}} \,.\, \alpha E b^2 d / \pi^2 D$ and its value is determined for increasing numbers of terms in the assumed deflexion mode, i.e.

$$w = \sum\sum a_{mn} \cos \frac{m\pi x}{a} \cos \frac{n\pi y}{b}, \quad (-a < x < a, \; -b < y < b) \tag{7.30}$$

The convergence of the results for increasing numbers of terms a_{mn} is shown below in Table 7.1.

TABLE 7.1. DEPENDENCE OF Φ ON NUMBER OF TERMS a_{mn}

Terms	Φ
a_{11}	6·35
a_{11}, a_{31}	5·65
a_{11}, a_{31}, a_{13}	5·30
a_{11}, a_{31}, a_{13}, a_{33}	5·39

It is obvious that terms a_{11} and a_{31} are the most important ones.

For further discussion on this and similar problems the reader should consult references 94, 118, 119 and 169. Thus, Miura(118) suggests that the most important temperature parameters affecting the stability of a symmetrically heated plate which is simply supported on *rigid edge members* are the average temperature rise of the plate over that of the edge members, and the non-uniformity parameter representing the difference between the maximum and minimum temperature rise. The local distribution is apparently of less significance which is seen by inspection of Figs. 7.2, 7.3 which are obtained from references 94 and 118. The thermal buckling parameter K_T is plotted against the plate aspect ratio for three, different forms of two-dimensional temperature distribution, viz.

$$(a)\quad T = T_{av}\left(A_{00} + A_{11}\sin\frac{\pi x}{2a}\sin\frac{\pi y}{2b}\right) \tag{7.31}$$

$$(b)\quad T = T_{av}\left(A_{00} + A_{11}\left\{1 - \left|\frac{x-a}{a}\right|\right\}\left\{1 - \left|\frac{y-b}{b}\right|\right\}\right) \tag{7.32}$$

$$(c)\quad T = T_{av}\left(A_{00} + A_{11}\left\{1 - \left[\frac{x-a}{a}\right]^2\right\}\left\{1 - \left[\frac{y-b}{b}\right]^2\right\}\right) \tag{7.33}$$

where

$$T_{av} = \frac{1}{4ab}\int_0^{2a}\int_0^{2b} T\,\mathrm{d}x\mathrm{d}y \tag{7.34}$$

The plates are of uniform thickness and the rigid edge members prevent any in-plane displacements.

Using the deflexion mode (7.28) results are presented in Fig. 7.2 for the distribution of (7.31), for either four terms, (a_{11}, a_{13}, a_{31}, a_{33}) $K_T^{(4)}$, or one term (a_{11}) $K_T^{(1)}$, 'and the difference in results is seen to be small. It should be noted that in Fig. 7.2 and Fig. 7.3, the

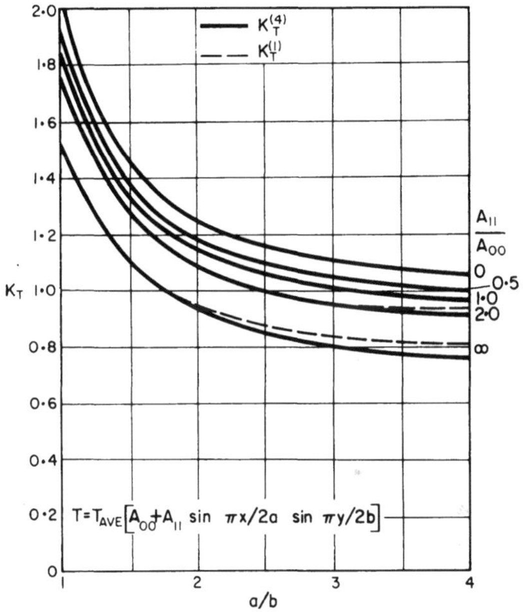

FIG. 7.2. Thermal buckling parameter for rectangular plate (rigid boundary) subjected to sinusoidal temperature distributions—Taken from reference 118. (*Aero. Res. Inst. Univ. Tokyo*).

curves for $A_{11}/A_{00} = 0$ may be obtained directly from isothermal elasticity for a plate with uniform stress acting and may be given by

$$K_T = 1 + b^2/a^2 \qquad (7.35)$$

where $$K_T = 48\,(1 + \nu)\alpha\, T_{av} \left(\frac{b}{d}\right)^2 \Big/ \pi^2$$

It may also be deduced from reference 118 that for a plate with *unrestrained* simply supported edges the shape of the local temperature distribution is of considerable significance whereas, of course, the term A_{00} does not enter the problem. This is seen in Fig. 7.4, also from reference 118, in which K_T is plotted against A_{11}/A_{00} for $a/b = 1$ and for a range of values of the edge restraint parameter Γ. If the

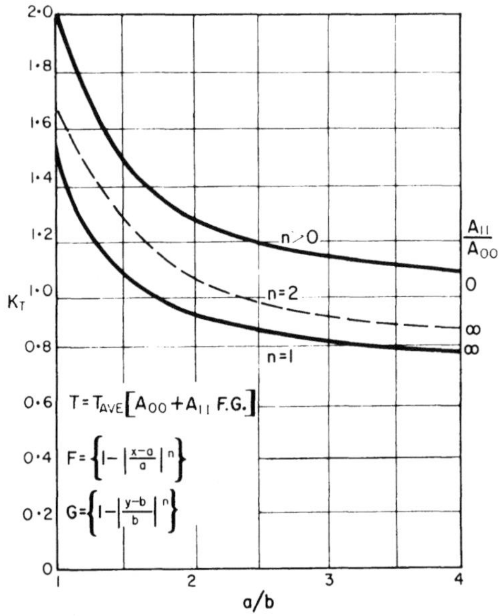

FIG. 7.3. Thermal buckling parameter for rectangular plate (rigid boundary) subjected to polynomial temperature distributions—Taken from reference 118. (*Aero. Res. Inst. Univ. Tokyo*).

edge members in the x and y directions have areas A_x, A_y respectively and Young's Moduli E_x, E_y then,

$$\Gamma = \frac{A_x E_x}{bdE} = \frac{A_y E_y}{adE} \qquad (7.36)$$

It is seen that when $\Gamma = \infty$ (rigid supports) the effect of the term A_{11}/A_{00} is at most 30 per cent. Decreasing Γ results in a rapid increase in K_T and in the effect of A_{11}/A_{00}. This is particularly noticeable when $\Gamma = 0$ (unrestrained plate).

7.3.5. Circular Plates

The Rayleigh–Ritz procedure outlined previously may also be applied to circular plates when the energy expressions are written in polar coordinates. Thus, for a uniform plate,

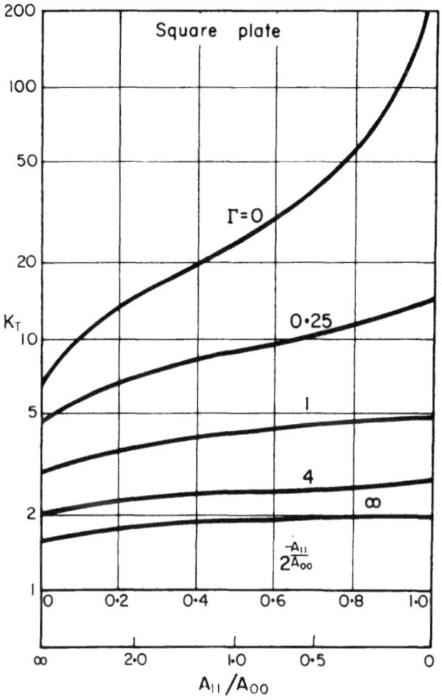

FIG. 7.4. Effects of edge restraint and temperature non-uniformity on thermal buckling—Taken from reference 118. (*Aero. Res. Inst. Univ. Tokyo*).

$$
\begin{aligned}
L = \frac{D}{2}\int\!\!\int_S \Bigg\{ & \left[\frac{\partial^2 w}{\partial r^2} + \frac{1}{r}\frac{\partial w}{\partial r} + \frac{1}{r^2}\frac{\partial^2 w}{\partial \theta^2}\right]^2 \\
& - 2(1-\nu)\left[\frac{\partial^2 w}{\partial r^2}\left(\frac{1}{r}\frac{\partial w}{\partial r} + \frac{1}{r^2}\frac{\partial^2 w}{\partial r^2}\right) - \left(\frac{1}{r^2}\frac{\partial w}{\partial \theta} - \frac{1}{r}\frac{\partial^2 w}{\partial r \partial \theta}\right)^2\right]\Bigg\}\, \mathrm{d}S \\
& + \frac{d}{2}\int\!\!\int_S \left\{\sigma_{rr}\left(\frac{\partial w}{\partial r}\right)^2 + \frac{\sigma_{\theta\theta}}{r^2}\left(\frac{\partial w}{\partial \theta}\right)^2 + \frac{2\sigma_{r\theta}}{r}\left(\frac{\partial w}{\partial r}\right)\left(\frac{\partial w}{\partial \theta}\right)\right\}\, \mathrm{d}S
\end{aligned}
\tag{7.37}
$$

There are several published papers in the literature dealing with this topic. A free circular plate under an axisymmetric stress system is considered in reference 171, and in reference 112 a clamped plate with a concentric circular hot spot is analyzed. The vibration characteristics of thermally stressed circular plates are examined in reference 17 and it is shown that as the critical temperature difference for buckling is approached the plate experiences a reduction of natural frequency to zero. This phenomenon of reduction in stiffness due to thermal stress is examined more closely in Section 7.6. Reference 19 summarizes a most thorough investigation into the thermal buckling problem for a complete range of edge boundary conditions and for two modes of heating, viz. over the centre and at the outside edge. In the former case the maximum compressive stresses are caused at the centre. The plate buckles into the form of a saucer and the degree of edge restraint has little effect on the plate behaviour. In the latter case the maximum compressive stresses are at the edge of the plate, the plate tends to buckle into a saddle shape, and the edge restraint is most important—it may inhibit buckling altogether.

In the analysis of reference 171, which showed good agreement with experiment the radial temperature distribution was represented by

$$T = T_0 + \Delta T\left(\frac{r}{b}\right)^n \tag{7.38}$$

and the axisymmetric stress distribution is determined using eqns. (3.23), (3.24), viz.

$$\left.\begin{aligned} \sigma_{rr} &= \frac{Ea\Delta T}{n+2}\left[1-\left(\frac{r}{b}\right)^n\right] \\ \sigma_{\theta\theta} &= \frac{Ea\Delta T}{n+2}\left[1-\left(\frac{r}{b}\right)^n(n+1)\right] \\ \sigma_{r\theta} &= 0 \end{aligned}\right\} \tag{7.39}$$

The assumed buckle displacement mode is

$$w = A\left(\frac{r}{b}\right)^2\left[1+\beta_1\left(\frac{r}{b}\right)^2+\beta_2\left(\frac{r}{b}\right)^4\right]\sin 2\theta \tag{7.40}$$

which corresponds to the experimentally observed saddle shape for edge heating conditions (see reference 19 and eqn. 7.38). The quantities β_1, β_2 are suitably chosen constants which make eqn. (7.40) satisfy the free edge boundary conditions (eqn. (4.32))

$$\left.\begin{aligned} M_r &= 0 \\ \frac{\partial M_r}{\partial r} - \frac{2\partial M_{r\theta}}{b\partial\theta} &= 0 \end{aligned}\right\} \tag{7.41}$$

Substitution into (7.37) of eqns. (7.39) and (7.40) yields the following result for the critical temperature difference to produce buckling

$$\Delta T_{cr} = 1{\cdot}607D/b^2dE\alpha F_{(n)} \tag{7.42}$$

where

$$F_{(n)} = \frac{n}{n+2}\left[\frac{1{\cdot}25}{n+4} + \frac{1{\cdot}25}{n+8}\beta_1^2 + \frac{1{\cdot}084}{n+12}\beta_2^2 + \frac{2{\cdot}67}{n+6}\beta_1 + \frac{2{\cdot}75}{n+8}\beta_2 + \frac{2{\cdot}4}{n+10}\beta_1\beta_2\right] \tag{7.43}$$

and $\beta_1 = -0{\cdot}279$, $\beta_2 = 0{\cdot}063$. The condition $d(\Delta T_{cr})/dn = 0$ gives $n = 2{\cdot}6$ as the most critical radial temperature distribution exponent, although ΔT_{cr} does not vary significantly for $1{\cdot}5 \leqslant n \leqslant 6$. It should be noted that the above analysis is based, essentially on a Rayleigh method; in other words the potential energy U has not been minimized with respect to arbitrary parameters in the deflexion mode for w, i.e. eqn. (7.40). A more detailed investigation would require a deflexion mode of the form[19]

$$w = \sum_{m=2,6,10} p_m \sin m\theta \tag{7.44}$$

in which p_m is a function of r only and contains an arbitrary coefficient with respect to which the potential energy is minimized (compare with (7.28)).

7.4. Post-Buckling (Large Deflexion) Analyses for Flat Plates

The large deflexion equations for plates have been presented in Section 4.2 and it is seen that the differential equation in w and the compatibility equation in ϕ_N must be satisfied simultaneously, where both equations are non-linear [eqns. (4.27) and (4.28)]. These equations are given below for convenience in a slightly different form:

$$D\nabla^4 w_1 - p + \frac{1}{1-\nu}\nabla^2 M_T - \left[\frac{\partial^2\phi_N}{\partial y^2}\frac{\partial^2 w}{\partial x^2} + \frac{\partial^2\phi_N}{\partial x^2}\frac{\partial^2 w}{\partial y^2} - \frac{2\partial^2\phi_N}{\partial x\partial y}\frac{\partial^2 w}{\partial x\partial y}\right] = 0 \tag{7.45}$$

$$\nabla^4\phi_N + \nabla^2 N_T - Ed\left[\left(\frac{\partial^2 w}{\partial x\partial y}\right)^2 - \frac{\partial^2 w}{\partial x^2}\frac{\partial^2 w}{\partial y^2}\right] + Ed\left[\left(\frac{\partial^2 w_0}{\partial x\partial y}\right)^2 - \frac{\partial^2 w_0}{\partial x^2}\frac{\partial^2 w_0}{\partial y^2}\right] = 0 \tag{7.46}$$

where $w = w_1 + w_0$.

One possible method of solution of these equations is to adopt an iterative procedure whereby an assumed form for w and the known form of w_0 are inserted into (7.46) which is then solved for ϕ_N so as to satisfy the boundary conditions on ϕ_N. The expressions for ϕ_N and w are then substituted into (7.45) and a new expression for w_1 found on solving (7.45). Thus a second form for w ($= w_1 + w_0$) can be inserted into (7.46) and the above procedure repeated until a sufficiently accurate result is obtained.

A less tedious method of solution would be to;

(1) assume a form for w which satisfies the kinematic boundary conditions e.g.

$$w = w\,(x, y, A_r, A_s, \ldots) \tag{7.47}$$

where A_r, A_s are initially undetermined parameters.

(2) Solve equation (7.46) exactly for ϕ_N in terms of A_r, A_s, so as to satisfy the boundary conditions on ϕ_N.

(3) Substitute these expressions for ϕ_N and w (7.47) into eqn. (7.45) and apply the Ritz–Galerkin procedure to (7.45), in order to determine the parameters A_r, A_s. Thus if (7.45) is written in the form $F_w(A_r, A_s, p, M_T, \ldots) = 0$, the A_r, A_s are determined from the simultaneous equations of the form

$$\iint F_w \cdot \frac{\partial w}{\partial A_r} \cdot \mathrm{d}x\mathrm{d}y = 0 \tag{7.48}$$

An alternative procedure[169] would be to assume that

$$\phi_N = \sum_{i=1}^{N} C_i \phi_{i(x,y)},$$

where the ϕ_i individually satisfy the boundary conditions and the coefficients C_i are determined as functions of the parameters A_r, A_s by the application of the Ritz–Galerkin procedure to eqn. (7.46). If (7.46) is written in the form $F_\phi(C_i, A_r, A_s, w_0, N_T, \ldots) = 0$, the C_i are determined from the simultaneous equations of the form

$$\iint F_\phi \cdot \phi_i \, dx dy = 0 \tag{7.49}$$

Step (3) above is then followed to determine the quantities, A_r, A_s.

Van Der Neut[169] suggests that the following approach is preferable. Equations (7.45) and (7.46) are replaced by the equilibrium equations in the x, y and z directions, and modes are assumed for a compatible system of displacement components u, v and w in the form of (7.47). Since such modes do not, in general, satisfy the equilibrium equations at all points in the plate, error functions are involved which may be visualized as " fictitious " pressures q_x, q_y, q_z (see eqn. (7.15)). The parameters A_r, A_s, etc., are then evaluated by application of the Ritz–Galerkin procedure. Thus

$$\iint \left\{ q_x \left(\frac{\partial u}{\partial A_r} \right) + q_y \left(\frac{\partial v}{\partial A_r} \right) + q_z \left(\frac{\partial w}{\partial A_r} \right) \right\} dx dy = 0 \ [r, s, \text{etc.}] \tag{7.50}$$

if the dynamic boundary conditions are satisfied by the assumed u, v and w. If the dynamic boundary conditions are not satisfied additional terms enter (7.50) representing the work done by the " additional loads " at the boundary (see eqn. (7.17) and reference 169).

The application of large deflexion theory to the problem of axisymmetric deflexions of solid circular plates is considered in reference 128. Solutions are obtained by solving the appropriate forms of (4.25) and (4.26) expressed in finite difference form. The plates are assumed to be simply supported on edges that are radially immovable, and results are presented for various loading conditions and temperature distributions. For a plate with a uniform temperature through the thickness ($M_T = 0$, $N_T \neq 0$) it is found that thermal

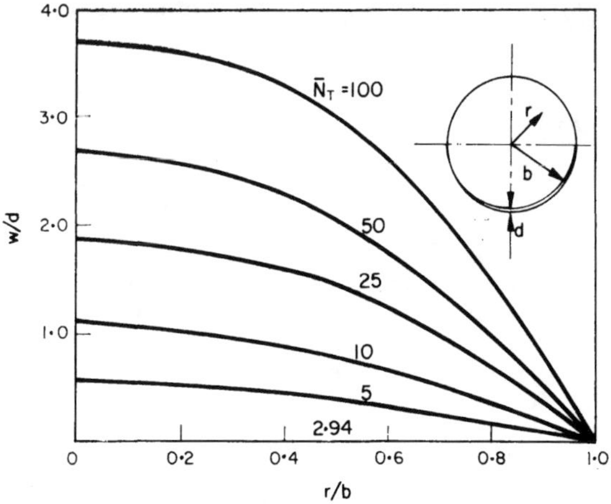

FIG. 7.5. Post-buckling behaviour of a circular plate; normal deflexions —Taken from reference 128. (*J. Aero-space Sci.*).

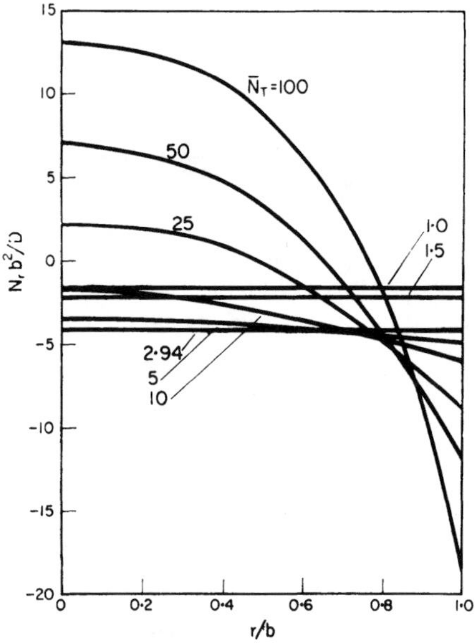

FIG. 7.6. Post-buckling behaviour of a circular plate; radial membrane forces—Taken from reference 128. (*J. Aero-space Sci.*).

buckling occurs when $\overline{N_T} = N_T b^2/D = 2{\cdot}94$. For lower values of $\overline{N_T}$ the plate is undeflected and the linear, elastic plane stress solution yields zero bending moments and uniform stresses over the planform. For higher values of $\overline{N_T}$, the lateral deflexions w increase monotonically with increasing $\overline{N_T}$. The stress resultants N_r and N_θ are compressive throughout the plate for low values of $\overline{N_T}$ but at higher

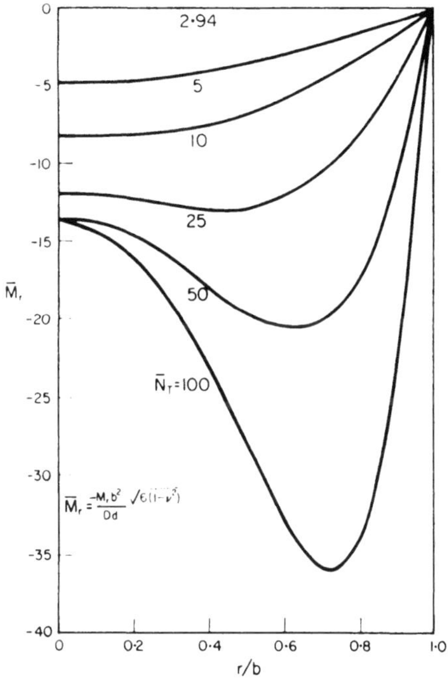

FIG. 7.7. Post-buckling behaviour of a circular plate; radial bending moment—Taken from reference 128. (*J. Aero-space Sci.*).

values of $\overline{N_T}$, the resultants are tensile in the central part of the plate. As $\overline{N_T}$ increases, the positions of maximum radial and tangential moments move from the centre of the plate towards the edge. These effects are illustrated in Figs. 7.5–7.7.

The post-buckling behaviour of simply supported rectangular plates has been examined in reference 57, for the temperature distribution of the form (7.29) and in references 44 and 156 for the temperature distribution of (7.33); reference 44 compares results with reference 156. A more general procedure allowing for various edge conditions is presented in reference 45.

7.5. Thermal Buckling of Circular Cylindrical Shells

7.5.1. Radial Temperature Variations

Thermal buckling due to a radial temperature variation through the thickness of a shell is unlikely unless certain kinds of external restraints are applied to the shell which can cause high average compressive stresses sufficient to produce instability.

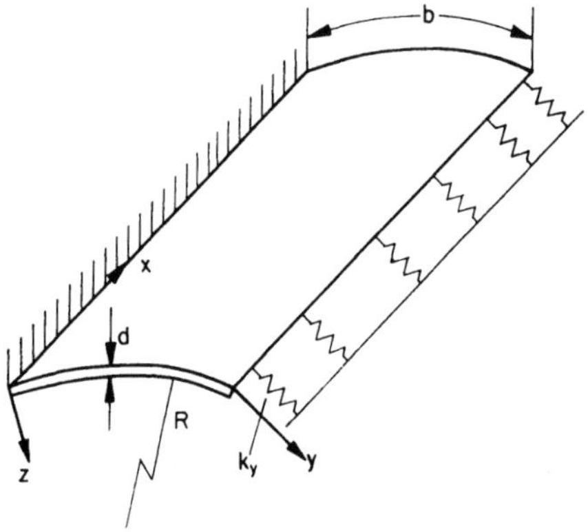

FIG. 7.8. Elastically restrained cylindrical panel—Taken from reference 97. (Reproduced from *Bull. Res. Counc.* 11C: 123 (1962), presently *Israel Journal of Technology*.)

Such a case is considered by Kornecki[97] and shown in Fig. 7.8. A shallow cylindrical panel of infinite length is hinged along its edges and a spring affords restraint against circumferential displacement. For a temperature distribution which is linear through the thickness, i.e. $T_{(z)} = T_0 + \Delta Tz/d$, the presence of the edge restraint induces hoop stresses in the curved panel and there is a possibility of buckling.

By using Marguerre's[109] large deflexion equations, (6.8) and (6.9), the governing equation for this problem reduces from (6.8) to

$$\frac{\partial^4 w}{\partial y^4} - \frac{d}{D}\sigma_{yy}\frac{\partial^2 w}{\partial y^2} = \frac{d\sigma_{yy}}{DR} \tag{7.51}$$

where σ_{yy} is a constant which is evaluated by considerations of the circumferential displacement in the presence of the spring and due to the temperature.

Kornecki presents detailed results for the case of rigid supports ($k_y = \infty$) and when the temperature distribution is

$$T_{(z)} = \frac{\Delta T}{d}\left(z + \frac{d}{2}\right),$$

i.e. no rise in temperature at the outer surface $z = -d/2$. It is found that the behaviour of the panel is mainly dependent on the ratio of its height (f) to its thickness (d), which for small angles of the projected angle, θ, in Fig. 7.8 may be written as $f/d = k/8$, where $k = b^2/Rd$. For $k \leqslant 1{\cdot}49$ no buckling occurs, but when $1{\cdot}49 < k < 1{\cdot}64$ (approximately) buckling can occur; such a panel buckles by a jump in the deflexion w and during the buckling the deflected shape is symmetrical. The critical value of ΔT is given by

$$\Delta T_{cr} = 2\left[\frac{\pi^2 D(1-\nu)}{db^2 E\alpha}\right] \tag{7.52}$$

where the term in the bracket corresponds to the uniform temperature rise necessary to produce buckling of an infinitely long hinged flat plate; and the critical compressive stress, σ_{yy}, is equal to the critical stress for a hinged flat strip (σ_F) given by $\pi^2 D/b^2 d$. It should be pointed out that there is increasing deflexion with increasing ΔT for all values of k less than $1{\cdot}49$ but true buckling does not occur, e.g. at no value of ΔT does the fundamental natural frequency become zero.

At larger values of k buckling can occur but only at larger values of the critical compressive stress σ_{yy} ($= 4\sigma_F$), again a jump takes place and during buckling the deflected shape contains a non-symmetrical part. A simple engineering formula for this case gives

$$\Delta T_{cr} = \frac{d}{\alpha R}\left[\frac{1 + \pi^2/4}{1+\nu}\right] \doteqdot \frac{2{\cdot}67}{\alpha}\frac{d}{R} \tag{7.53}$$

which is valid for $f/d \geqslant 15$, assuming $\nu = 0{\cdot}3$.

7.5.2. Axial Temperature Variations

Cylindrical shells may experience compressive hoop stresses when either there is an axial temperature variation or, if the temperature rise is uniform, restraint is provided by cooler elements against the uniform expansion of the shell. In this latter case the discontinuity stresses (Section 6.6) set up near the junction of an axisymmetrically heated cylindrical shell and a cooler stiffening ring or bulkhead may be sufficiently high to cause shell buckling.

Hoff[70] first investigated the stability of a simply supported cylindrical shell subjected to a uniform temperature rise. Infinite trigonometric series are used to represent the radial deformation of the shell and the axial distribution of hoop stress. Donnell's simplified eighth-order small deflexion theory[28] is used to obtain a solution in the form of an infinite determinant which can be truncated to give a solution to any desired degree of accuracy. The analysis has an almost identical form to that of Section 7.3.4.

For clamped ended shells such a direct analysis is not possible and a solution of Donnell's equation requires the application of the Galerkin procedure.[88,175] Anderson[2] uses the modified equation of equilibrium proposed by Batdorf[10] for both clamped and simply supported ends, whilst Sunakawa[157] develops yet another fourth-order equation. These various equations may be written as follows,

$$D\nabla^8 w + \frac{Ed}{R^2}\left(\frac{\partial^4 w}{\partial x^4}\right) + d\nabla^4\left(\sigma_{yy}\frac{\partial^2 w}{\partial y^2}\right) = 0 \quad \text{(Donnell}^{(28)}\text{)} \tag{7.54}$$

$$D\nabla^4 w + \frac{Ed}{R^2}\nabla^{-4}\left(\frac{\partial^4 w}{\partial x^4}\right) + d\sigma_{yy}\frac{\partial^2 w}{\partial y^2} = 0 \quad \text{(Batdorf}^{(10)}\text{)} \tag{7.55}$$

$$D\left[\nabla^4 w + \frac{2}{R^2}\left(\frac{\partial^2 w}{\partial y^2} + \nu\frac{\partial^2 w}{\partial x^2}\right) + \frac{w}{R^4}\right]$$

$$+ d\sigma_{yy}\left(\frac{\partial^2 w}{\partial y^2} + \frac{w}{R^2}\right) + \frac{d}{R}\sigma'_{yy} = 0 \quad \text{(Sunakawa}^{(157)}\text{)} \tag{7.56}$$

and all require the use of the Galerkin procedure when applied to clamped shells. Note that σ_{yy} is the pre-buckling compressive hoop stress and σ'_{yy} is the change in σ_{yy} due to buckling.

When infinite trigonometric series are used to represent w and σ_{yy} as in references 2 and 70, e.g.

$$w = \sin\frac{ny}{R}\sum_{p=1}^{\infty} a_p \sin\frac{\pi x}{L}\sin\frac{p\pi x}{L} \quad \text{(clamped ends)} \tag{7.57}$$

or

$$w = \sin \frac{ny}{R} \sum_{p=1}^{\infty} a_p \sin \frac{p\pi x}{L} \quad \text{(simply supported ends)} \tag{7.58}$$

and

$$\sigma_{yy} = EB \sum_{m=0}^{\infty} S_m \cos \frac{m\pi x}{L}, \quad (B \text{ is an eigenvalue}) \tag{7.59}$$

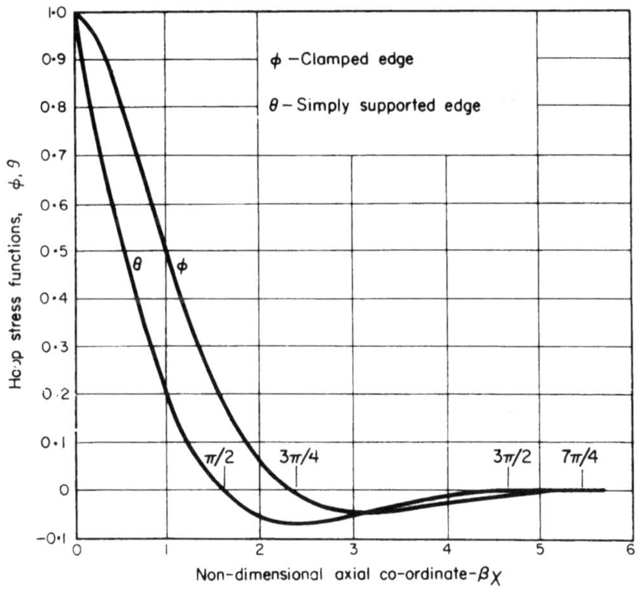

FIG. 7.9. Hoop stress functions for a circular cylindrical shell.

it is found that many terms are required in each series to adequately represent the actual conditions in the shell. This is so because the discontinuity stresses decrease rapidly away from the shell-bulkhead joint at $x = 0$ and the problem is essentially one of local buckling. Thus, for a uniformly heated shell attached to rigid, non-expanding rings or bulkheads the exact, compressive hoop stress distributions shown in Fig. 7.9 are

$$\sigma_{yy}/E\alpha T = \phi = e^{-\beta x}(\cos\beta x + \sin\beta x) \quad \text{(clamped)} \tag{7.60}$$

$$\sigma_{yy}/E\alpha T = \theta = e^{-\beta x}\cos\beta x \quad \text{(simply supported)} \tag{7.61}$$

and it is obvious that many terms S_m would be required in eqn. (7.59), with $B = \alpha T_{cr}$, for adequate representation of σ_{yy} over the entire length of shell. Because of this it has been suggested[83] that a more important parameter in this problem than the shell length (L) would be the length of shell near to both joints for which the hoop stresses are compressive i.e. from Fig. 7.9, $L_\beta = 3\pi/2\beta$ for clamped ends and $L_\beta = \pi/\beta$ for simply supported ends. This concept is used in references 88 and 89 for clamped shells, and also in reference 65 where an energy method of solution is presented which only considers a region close to the shell-bulkhead joint. The approach of reference 157 is probably the best in that σ_{yy} is retained in its exact form (7.60), and the assumed buckle mode, whilst it is applicable to the entire length of shell, puts most emphasis on the region near the joint, viz.

$$w = A\sin\frac{ny}{R}e^{-\beta x}\left(1 - \cos\frac{\pi x}{l}\right) \tag{7.62}$$

where A is an arbitrary coefficient and l, n are half the axial wavelength and the number of circumferential waves in the buckle mode respectively. Both l and n are determined in the process of minimizing T_{cr}, the critical shell temperature rise. An alternative form for w (reference 89) applicable over the length of shell NL_β ($< L$) would be, for clamped shells,

$$w = \sin\frac{ny}{R}\sum_{p=1}^{\infty} a_p \sin\frac{2\pi x}{NL_\beta}\sin\frac{2p\pi x}{NL_\beta} \tag{7.63}$$

which satisfies the conditions $w = \partial w/\partial x = 0$ at $x = 0$, $NL_\beta/2$. In this case T_{cr} is minimized with respect to N and n.

Whilst any one of the three governing equations (7.54)–(7.56) might be used in any particular problem it would seem that the Donnell eighth-order equation (7.54) is suspect when used with the

Galerkin procedure for clamped ended shells. This has been pointed out by Batdorf[10] and Anderson[3] and further relevant comments are given by Singer[152]. Thus although it is termed an equilibrium equation, (7.54) has been derived by differentiation from the original fourth-order equation for radial equilibrium to produce the more convenient form of (7.54). Singer[152] states that agreement with the Rayleigh–Ritz method (and hence certainty of an upper bound solution for instability problems) is ensured only when the Galerkin procedure is applied to the equilibrium equation of the problem that resulted from the variation of the total potential energy. That the Donnell equation may be used for simply supported shells is therefore fortuitous since for a mode of the form of (7.58) there is no essential difference between the functions $\nabla^8 w$ and $\nabla^4 w$. This is not so for clamped shells with modes of the form of (7.57).

Results available from the literature show that the critical uniform temperature rise, T_{cr}, may be written as (compare with (7.53))

$$T_{cr} = \frac{K}{\alpha}\frac{d}{R} \tag{7.64}$$

where, for a single shell with simply supported ends $K = 5{\cdot}3$[70,88], for a single or multibay shell with clamped ends $K = 3{\cdot}76$[2,3,157] and for a multibay shell with simply supported ends $K = 2{\cdot}1$.[2,3] The fact that the clamped shell is more prone to buckling than one with simply supported ends (3.76 cf. 5.3) may be reasoned from Fig. 7.9 where it is seen that although clamping gives a more rigid support to the shell, the compressive hoop stresses are larger and act over a 50 per cent greater shell length. For a multibay shell consisting of a uniform shell simply supported or clamped on several rings or bulkheads, the compressive hoop stresses must have the form of (7.60) and in this case the simply supported shell must be more prone to buckling than the clamped shell (2.1 cf. 3.76).

For a non-uniform axial temperature distribution in a shell supported on a flexible, expanding bulkhead, the methods outlined previously are still applicable provided the shell hoop stress distribution is correctly represented in the analysis. This has been done in reference 88 which also contained the photograph of the thermal buckle pattern shown in Fig. 7.10.

Corresponding results for conical shells are given in reference 151.

7.5.3. Circumferential Temperature Variations

Cylindrical shells should certainly be analyzed for this type of temperature distribution since the conditions which are set up are

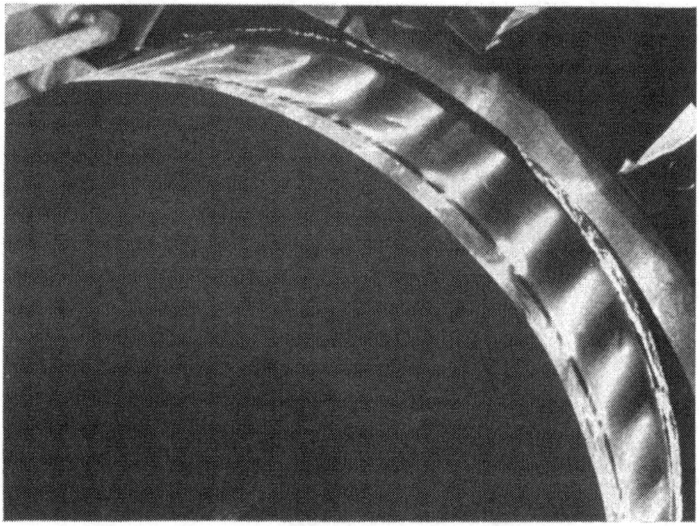

FIG. 7.10. Thermal buckles due to discontinuity stresses in a thin circular cylindrical shell.

very similar to those which are known to cause buckling in flat or near-flat plates (see Section 7.3.4).

A detailed analysis on this subject for a thin isotropic cylindrical shell, simply supported at the ends, is given in reference 1, which uses the Donnell eighth-order equation,

$$D\nabla^8 w + \frac{Ed}{R^2}\frac{\partial^4 w}{\partial x^4} + d\nabla^4 \left\{\sigma_{xx}\frac{\partial^2 w}{\partial x^2}\right\} = 0 \qquad (7.65)$$

The compressive axial stress σ_{xx} is assumed to be constant in the radial and axial directions, and its circumferential variation is represented by an infinite Fourier series of the form,

$$\sigma_{xx} = EB \sum_{p=0}^{\infty} S_p \cos \frac{py}{R} \text{ (}B\text{ is an eigenvalue)} \tag{7.66}$$

As in reference 70 and in Section 7.3.4 an infinite determinant is obtained which may be truncated to give any required degree of accuracy to the buckling criterion. The assumed buckle mode for w is

$$w = \sin\left(\frac{m\pi x}{L}\right) \sum_{n=0}^{\infty} a_n \cos \frac{ny}{R} \tag{7.67}$$

where m and $2n$ are the number of half-waves along the generator of the shell and around the circumference respectively. The shell is assumed to be long compared with the half-wavelength in the x direction. By substituting (7.66) and (7.67) into (7.65) and after some manipulation the following infinite determinant of the coefficients of a_n is obtained

$$\begin{vmatrix} \frac{1}{2}(U_0 r - 2S_0), & -S_1, & -S_2, & -S_3, & \ldots \\ -S_1, & U_1 r - 2S_0 - S_2, & -S_1 - S_3, & -S_2 - S_4, & \ldots \\ -S_2, & -S_1 - S_3, & U_2 r - 2S_0 - S_4, & -S_1 - S_5, & \ldots \\ -S_3, & -S_2 - S_4, & -S_1 - S_5, & U_3 r - 2S_0 - S_6, & \ldots \\ \cdot & \cdot & \cdot & \cdot & \\ \cdot & \cdot & \cdot & \cdot & \end{vmatrix} = 0 \tag{7.68}$$

where $r = \dfrac{1}{B}$, and $U_n = \left\{\left[\left(\dfrac{m\pi R}{L}\right)^2 + n^2\right]^2 \Big/ 2\beta^4 \left(\dfrac{m\pi R}{L}\right)^2\right\}$

$$+ \left\{2\left(\frac{m\pi R}{L}\right)^2 \Big/ \left[\left(\frac{m\pi R}{L}\right)^2 + n^2\right]^2\right\}$$

To solve a given problem, the parameters S_p which characterize the axial stress distribution are inserted and the quantities U_n determined for a specific value of m. The determinant is then truncated and expanded into an equation yielding roots of r as a function of the assumed value of m. Since it is required to find the minimum value of B, the above procedure is repeated for various m until the maximum root of r is obtained.

Since the analysis is based on small deflexion theory the results obtained corresponding to uniform axial compression ($S_0 = 1$, $S_p = 0$ for $p \neq 0$) and pure bending ($S_1 = 1$, $S_p = 0$ for $p \neq 1$) agree with those from previous isothermal analyses. However the most interesting result was that found for arbitrary S_p; viz. that with little error, the critical value of a thermal and/or mechanical axial stress distribution around the periphery of a long, thin-walled cylindrical shell is reached when the maximum compressive stress at any point is equal to the classical critical stress for the shell subjected to uniform axial compression. Since there is considerable scatter in the results for thin shells subjected to uniform axial compression, due to initial irregularities etc., the adoption of such results for the present thermal problem must be treated carefully.

However, although the maximum axial thermal stress at buckling may be known it is another problem to determine the temperature distribution at which this stress occurs. For long unstiffened shells elementary theory is probably adequate (e.g. that of Section 5.2) except near the ends where the axial stresses must decay to zero, but for ring-stiffened shells the elementary theory does not account for radial deflexions of the rings or cylinder wall. These deflexions cannot be neglected for typical dimensions of ring-stiffened cylinders and the elementary theory must be modified accordingly. An approach which has been justified experimentally is given in reference 4.

7.6. Thermal Buckling of Thin Wings

7.6.1. General Instability

It can be shown[96, 105] that one of the many problems arising from aerodynamic heating of thin wings in high speed flight is that the middle surface thermal stresses may significantly reduce the overall stiffness of the wing and produce overall buckling, e.g. in torsion or spanwise flexure.

For representative wing cross-sectional shapes, whether solid or hollow, aerodynamic heating usually causes spanwise thermal stresses which are compressive at the leading and trailing edges, and tensile at the mid-chord. When such a wing is twisted these middle surface stresses have a resultant torque, see Fig. 7.11, which acts in the same sense as the twist and so produces an apparent reduction in torsional stiffness. Small deflexion theory[169] for this problem results in the following formula for the apparent torsional stiffness

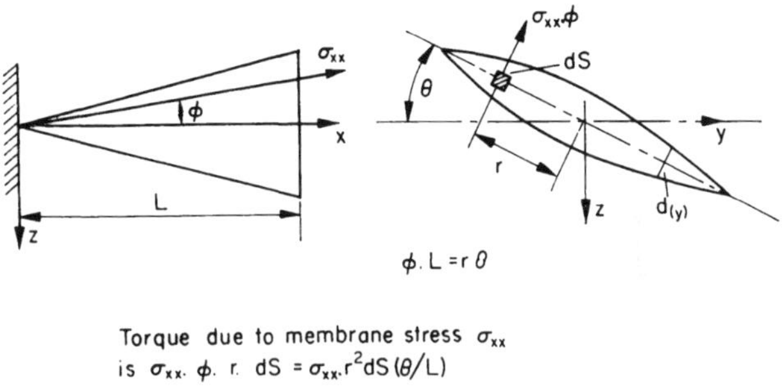

FIG. 7.11. Torque due to membrane stresses in a thin twisted wing.

of a wing (or plate) with a normal stress distribution over the cross-section,

$$\overline{GC} = GC + \iint_s \sigma_{xx} r^2 \mathrm{d}S \tag{7.69}$$

where $\overline{GC}$ is the apparent value and GC is the nominal value for the section according to Saint-Venant, calculated with due consideration of the effects of temperature on G the shear modulus; $\mathrm{d}S$ is an infinitesimal area and r is the distance of $\mathrm{d}S$ from some reference axis normal to the plane of the cross-section. The integral above may be evaluated with respect to any such axis. It follows from (7.69) that for a wing of thickness $d_{(y)}$, if the normal thermal stress $\sigma_{xx} = Ea\Delta T f_{(y,z)}$, thermal buckling occurs at the value of ΔT_{cr} given by

$$\Delta T_{cr} = -\, GC/Ea \iint f_{(y,z)}\, r^2 \mathrm{d}S \tag{7.70}$$

which for a very thin wing may often be approximated to

$$\Delta T_{cr} = -\, GC/Ea \int f_{(y)}\, y^2 d_{(y)}\, \mathrm{d}y \tag{7.71}$$

When $\Delta T < \Delta T_{cr}$, eqn. (7.69) may be rewritten (if f is invariant with time) as,

$$\overline{GC} = GC\,[1 - (\Delta T/\Delta T_{cr})] \tag{7.72}$$

Since the function $f_{(y,z)}$ corresponds to the particular temperature distribution which varies with time (as also does ΔT) it is not obvious that buckling is most likely when ΔT is a maximum, and the complete heating phase should be analyzed for possible buckling (eqn. (7.72) should therefore be used with caution). In practice buckling should not occur in a good design, but the reduction of stiffness should be evaluated because of its effect on the wing's aeroelastic behaviour.

Similarly if the same wing is bent in the spanwise direction, distortion of the cross-section occurs due to the anticlastic effect and the radial component of the middle surface stresses. Due to this distortion the middle surface stresses may have a resultant moment acting on the cross-section in the same sense as the applied moment thus producing an apparent reduction in flexural stiffness.[(104)]

The combined torsional and flexural problem is studied by Mansfield in reference 105 using a large deflexion analysis for a class of wings of infinite aspect ratio. With regard to the effect of finite aspect ratio, Mansfield concludes[(104)] that for uniform wings with chordwise temperature variations only the decay of the spanwise middle surface stresses towards the wing tip is important for aspect ratios less than about 2. In reference 104, it is assumed that the magnitude of the spanwise middle surface stresses varies as

$$F_{(x)} = \left[1 - \frac{\cosh(x/\psi b)}{\cosh(a/\psi b)}\right]$$

where x is zero at the centre line, $2a$ is the overall wing span, ψ is a decay parameter and $2b$ is the wing chord.

Thus, $\sigma_{xx} = E\alpha\Delta T f_{(y,z)}\,F_{(x)}$ and ψ is found from the condition that the strain energy is a minimum with respect to ψ. The effect of finite aspect ratio (viz. the variation of $F_{(x)}$ to zero at $x = a$) is to increase the value of ΔT_{cr} for thermal buckling compared with eqns. (7.70) or (7.71) which are essentially for a wing of infinite aspect ratio with $F_{(x)} = 1$.

The large deflexion analysis of reference 105 enables relationships to be obtained between bending moment, torque, spanwise curvature and twist as a function of the middle surface stresses. The apparent flexural and torsional stiffnesses are seen to be functions of both the spanwise curvature and twist, but a more significant result is that increasing the deflexions (curvature or twist) has a stabilizing influence on the apparent stiffness and on the buckling temperature. A brief note by Bisplinghoff[13] compares theoretical and experimental results for finitely twisted flat plates subjected to a chordwise temperature distribution described by an even function $T_{(y)}$. Neglecting any warping restraint the torque M_T is related to the twist per unit length θ by (from (7.69))

$$M_T = \theta \, [4bD \, (1 - \nu) + d \int_{-b}^{b} \sigma_{xx} y^2 \mathrm{d}y] \tag{7.73}$$

where $D = Ed^3/12(1 - \nu^2)$, $2b$ is the plate chord and d is the plate thickness. For finite deflexions,

$$\sigma_{xx} = \frac{E\theta^2}{6}\left[3y^2 - b^2\right] + \frac{E\alpha}{2b}\left[\int_{-b}^{b} T\mathrm{d}y - 2bT\right] \tag{7.74}$$

thus if $T = T_0 + \Delta T g_{(y)}$ eqn. (7.73) becomes

$$M_T/4bD\,(1 - \nu)\theta = 1 - \frac{\Delta T}{\Delta T_{\mathrm{cr}}} + \frac{4}{15}\,(1 + \nu)\left(\frac{b^4\theta^2}{d^2}\right)$$

$$= 1 - \frac{\Delta T}{\Delta T_{\mathrm{cr}}} + \frac{4}{45(1 - \nu)}\,\Omega^2 \tag{7.75}$$

where ΔT_{cr} is the temperature difference required to produce buckling at zero twist, viz.

$$\Delta T_{\mathrm{cr}} = bd^2/3(1 + \nu)\alpha\left[\int_{-b}^{b} gy^2\mathrm{d}y - \frac{b^2}{3}\int_{-b}^{b} g\,\mathrm{d}y\right] \tag{7.76}$$

and Ω is the twist rate parameter $b^2\sqrt{(Ed\theta^2/4D)}$. In Fig. 7.12 (reference 13) a comparison is shown between eqn. (7.75) and experimental results for a flat plate twisted under the conditions of $\Delta T/\Delta T_{cr} = 0$, 0·4 and 1·0; the temperature distribution across the chord is tent-like, i.e. $g_{(y)} = |y/b.|$ Although reasonable correlation was obtained between eqn. (7.76) and the experimental value

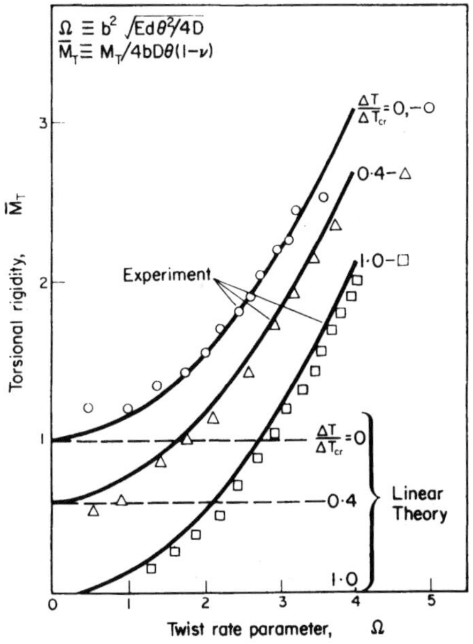

Fig. 7.12. Torsional rigidity of a heated and twisted thin flat plate—Taken from reference 13. (*J. Aero-space Sci.*).

of ΔT_{cr}, this latter value is used in plotting the solid curves of Fig. 7.12.

Most analyses on wing buckling problems are concerned with symmetric sections, e.g. biconvex, diamond sections, etc. Certain wing sections which have been proposed for hypersonic applications, e.g. single wedge sections etc., may, because of their asymmetry, experience considerable spanwise distortion as the spanwise thermal stresses increase, even in the absence of a spanwise bending moment.

7.6.2. Local Instability

Another possible mode of instability for thin wing sections is that localized mode in which the thin leading and/or trailing edge sections buckle in the spanwise direction before overall wing buckling occurs. Mansfield(106) considers this problem and deduces a criterion whereby the possibility of local instability occurring before overall flexural or torsional instability may be predicted. In a later note(108) Mansfield shows that complete loss of torsional stiffness cannot occur before complete loss of flexural stiffness and only when $\phi_N \propto D$ do both stiffnesses vanish simultaneously.

The compressive stress resultant $N_x = \mathrm{d}^2\phi_N/\mathrm{d}y^2$ where ϕ_N is the force function satisfying the boundary conditions $\phi_N = \mathrm{d}\phi_N/\mathrm{d}y = 0$ at $y = \pm b$. At the onset of buckling let $\phi_N = \gamma\phi_0$ where ϕ_0 is fixed and γ is a constant of proportionality. By equating the middle surface strain energy to the bending strain energy at buckling it may be shown(106) that

$$\gamma = \frac{\iint D\left[(\nabla^2 w)^2 + 2(1-\nu)\left\{\left(\frac{\partial^2 w}{\partial x \partial y}\right)^2 - \frac{\partial^2 w}{\partial x^2}\frac{\partial^2 w}{\partial y^2}\right\}\right]\mathrm{d}x\mathrm{d}y}{2\iint \phi_0 \left\{\left(\frac{\partial^2 w}{\partial x \partial y}\right)^2 - \frac{\partial^2 w}{\partial x^2}\frac{\partial^2 w}{\partial y^2}\right\}\mathrm{d}x\mathrm{d}y} \tag{7.77}$$

An exact minimization of this expression is obviously not possible in general, but if $\phi_0 = D$ a simple result is found, viz. the minimum (positive) value of γ occurs when

$$\nabla^2 w = 0 \tag{7.78}$$

and is given by

$$\gamma^{(+)} = (1-\nu)$$

i.e.

$$\phi_N^{(+)} = (1-\nu)D \tag{7.79}$$

Modal forms which satisfy (7.78) include torsional and flexural buckling modes as well as the leading edge local buckling mode which is given typically by

$$w \propto \begin{pmatrix}\sin\\ \cos\end{pmatrix} (x/\chi) \exp(\pm y/\chi) \tag{7.80}$$

In this case when $\phi_N \propto D$, buckling of the leading edge occurs simultaneously with overall flexural and torsional buckling at a critical buckling stress distribution given by

$$\sigma_{xx}^{(+)} = N_x^{(+)}/d = \left(\frac{1-\nu}{d}\right)\frac{\mathrm{d}^2 D}{\mathrm{d}y^2} \tag{7.81}$$

for a solid wing, or

$$\sigma_{xx}^{(+)} = G\left\{(d')^2 + \frac{d}{2}(d'')\right\} \tag{7.82}$$

where primes denote differentiation with respect to y the chordwise coordinate. The corresponding result for a thin-walled wing of constant skin thickness t is

$$\sigma_{xx}^{(+)} = N_x^{(+)}/2t = G\left\{(d')^2 + d(d'')\right\} \tag{7.83}$$

if only the skins carry end load. Thus it can be shown that for solid and thin-walled wings with sharp leading edges ($d = 0$ at $y = \pm b$) leading edge buckling is more likely than overall buckling when σ_{xx} at the leading edge exceeds $G(d')^2_{LE}$. This is the simple criterion obtained by Mansfield[106] and which may be written as

$$(\sigma_{xx})_{LE} > G\,\mu^2 \ngtr E\mu^2/2(1+\nu) \tag{7.84}$$

where $\mu = (d')_{y=\pm b}$ is the total, enclosed leading edge angle.

As an example[106] consider a 3 per cent solid steel wing of diamond (or double-wedge) section having a chordwise temperature distribution of

$$T = T_0 + \Delta T\left|\frac{y}{b}\right|^n, \quad n = 0, 1, 2, \ldots \tag{7.85}$$

The spanwise thermal stress distribution is given by (eqn. (5.16)),

$$\sigma_{xx} = -E\alpha\Delta T\left[\frac{2}{(n+1)(n+2)} - \left|\frac{y}{b}\right|^n\right] \tag{7.86}$$

which at the leading or trailing edge becomes

$$(\sigma_{xx})_{LE} = E\alpha\Delta T\,.\,n(n+3)/(n+1)(n+2) \tag{7.87}$$

Therefore local instability will occur when

$$\Delta T > \frac{\mu^2}{2(1+\nu)\alpha}\cdot\frac{(n+1)(n+2)}{n(n+3)} \tag{7.88}$$

which for the example considered and $n = 2$ gives the result that the particular aerofoil suffers local instability when $\Delta T > 144°C$. It can be shown[106] that the torsional and flexural rigidities at this temperature difference are still 0·72 and 0·83, respectively, of their orginal (unheated) values.

Once local buckling occurs it is possible that any further increases in the temperature difference ΔT only cause a growth in the local

FIG. 7.13. Thermal buckles at the leading edge of a thin wedge-shaped plate.

buckling mode (i.e. χ in (7.80) increases) but the middle surface stresses may not increase. One might imagine that structurally this effect is beneficial since it should preclude the possibility of overall thermal buckling. Even if this is so, leading edge waviness would probably be unacceptable for aerodynamic reasons. To illustrate the order of magnitude which these leading edge buckles might have, Fig. 7.13 shows an experimental result for a thin diamond section wing.*

The corresponding analysis for a built-up wing section Fig. 7.14 is given by McKenzie[113]. The leading edge buckling parameter

* Figure 7.13 is taken from an unpublished thesis submitted by D. S. Mishra to the College of Aeronautics, Cranfield, U.K. in June, 1962, entitled "The Effect of Leading Edge Thermal Buckling on Wing Stiffness".

is defined by $\lambda = \sqrt{(N_x e^2/D)}$ where N_x is the spanwise compressive thermal stress resultant ($N_x = t_s \sigma_{xx}$) in each skin which is assumed constant over length e. The corresponding stress level in the solid edge fillet is $k\ \sigma_{xx}$ where the parameter k may be less than or greater than 1 and even negative. Results are obtained in reference 113 for λ for a range of values of k between $-1, 0, 1$ and 2 and for the range of

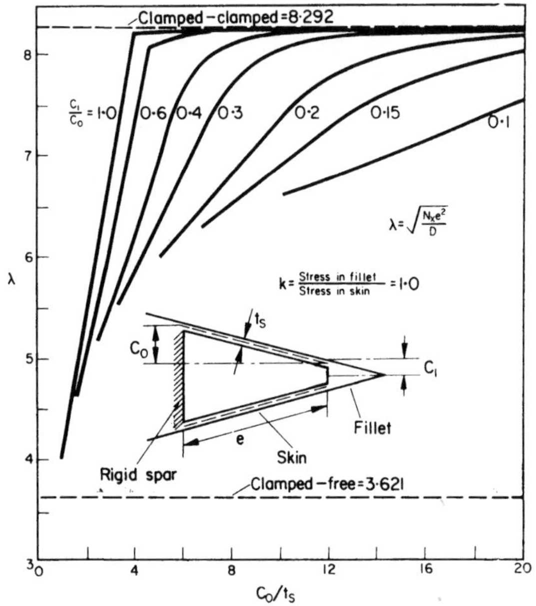

FIG. 7.14. Thermal buckling parameter for leading edge of a built-up wing—Taken from reference 113. (Crown Copyright: Reproduced by Permission of H.M. Stationery Office).

geometrical parameters $0{\cdot}1 < C_1/C_0 < 1$, $0 < C_0/t_s < 20$ [Note. $C_1 \geqslant t_s$]. The results indicate little variation of λ with k for the smaller values of C_1/C_0 but more variation as $C_1/C_0 \rightarrow 1$. Fig. 7.14, reproduced from reference 113 for $k = 1{\cdot}0$, illustrates a typical set of results. The two dotted lines refer to the end conditions assumed at the skin–spar joint and at the skin–fillet junction and represent extreme conditions for the skin stability. The analysis considers the skin to be clamped to both the spar and the fillet: the spar is rigid but the fillet has finite torsional and flexural stiffness.

CHAPTER 8

Sundry Design Problems

8.1. Temperature-Dependent Material Properties

The previous chapters have been primarily concerned with analyses in which it is assumed that the material properties are independent of temperature and the thermal stresses are in the elastic portion of the material stress–strain curve. Unfortunately, however, most structural materials are significantly temperature dependent in their elastic modulus so that it is necessary to modify the previously derived formulae to the more general problem.

A treatment on these lines for thin plates is given in reference 127 with the result that when E is varying with temperature, which itself varies through the plate thickness, the following definitions apply (assuming negligible variation in Poisson's ratio ν),

$$\left.\begin{aligned} \overline{Ed} &= \int_d E\mathrm{d}z,\ \overline{D} = \frac{1}{1-\nu^2}\int_d Ez^2\mathrm{d}z \\ N_T &= \int_d E\alpha T\mathrm{d}z,\ M_T = \int_d E\alpha Tz\mathrm{d}z \end{aligned}\right\} \tag{8.1}$$

These quantities should be compared with their corresponding forms in Chapter 4 (eqns. (4.10) and (4.11)). Temperature dependent coefficients of expansion cause no difficulty as it is sufficient only to retain the product αT as a parameter.

The effect of variations of Young's Modulus, E, with temperature on the thermal buckling of plates has been considered by Van Der Neut[169]. It is suggested that,

(*a*) When the compressive stresses are below the proportional limit E is assumed constant and having the lowest value appropriate to any temperature in the plate.

(*b*) When the compressive stresses are in the plastic range E is replaced by the lowest secant modulus appropriate to the temperature and stress level at any point in the plate, and assumed constant. These artifices need only apply to the actual buckling analysis (as a more refined approach is possible for the initial thermal stress analysis) and are essentially conservative.

The material properties to be considered which most influence (or are most influenced by) high temperature are, respectively, the emissivity, thermal conductivity, specific heat and specific gravity,

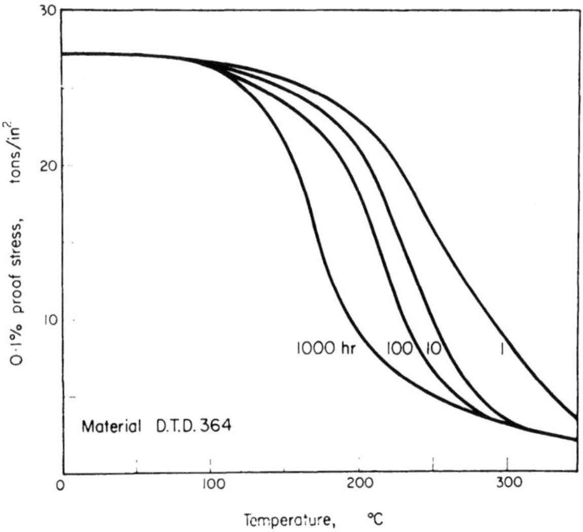

FIG. 8.1. Typical variation of 0·1 per cent proof stress with temperature as a function of time at temperature.

and the values of the mechanical strengths, elastic moduli and coefficients of thermal expansion. It is known that any finite exposure time at a sufficiently high temperature will cause a marked reduction in mechanical properties. Figure 8.1 shows how the 0·1 per cent proof stress of a typical aluminium alloy D.T.D.364 is a function both of temperature and exposure time. Consequently it must be emphasized that the material properties assumed in any analysis must be realistic for the exposure times and strain rates anticipated in the particular application, e.g. 1000 hr exposure data are too conservative for use in a guided missile design, but may be unconservative for use in a supersonic transport aircraft design. To indicate the significance of heating rates on material properties Fig. 8.2 has been prepared from data available in the literature.

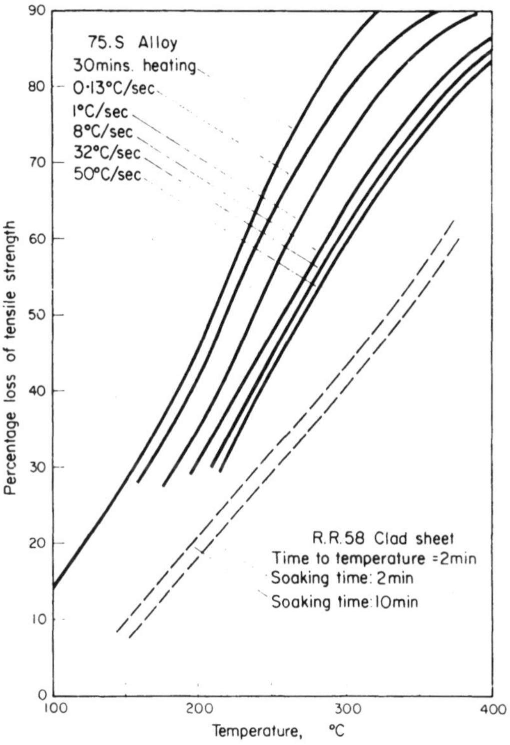

FIG. 8.2. Reduction of tensile strength at elevated temperature as a function of heating rate and time at temperature.

The alloy 75S is equivalent to D.T.D.687 and R.R.58 is a Hiduminium alloy classified as D.T.D.5070 when in clad sheet form. The results for 75S indicate that very fast heating rates are extremely beneficial and they suggest that the percentage loss of tensile strength of clad R.R.58 at the same heating rates should be quite low—particularly as the plotted R.R.58 values correspond to a mean heating rate of about 2 °C/sec. To indicate the orders of some of the above-mentioned effects Tables 3 and 4 in the Appendix give illustrative values of material and physical properties at various temperatures.

Another point to be noted is that in general the thermal stresses occur simultaneously with those due to applied external loads. If it can be assumed that the material remains elastic, the individual stresses and of course the strains may be added. However, if the stresses either individually or together exceed the limit of proportionality it is no longer justifiable to add either stresses or strains.

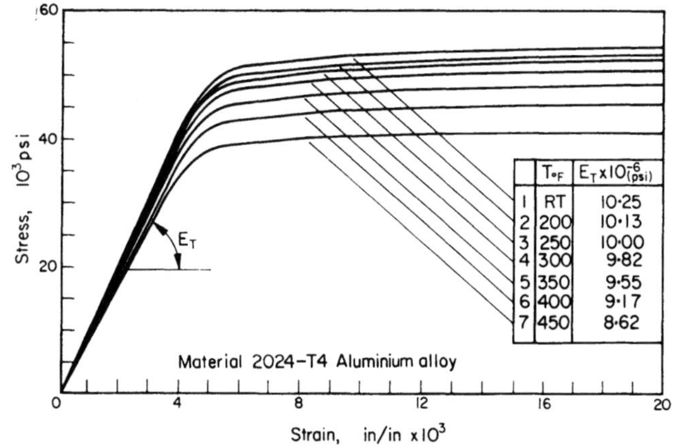

FIG. 8.3. Typical stress–strain curves at various temperatures—Taken from reference 153. (*Soc. Auto. Engrs.*)

In fact, the stresses and strains may depend on the particular order in which the applied and thermal stresses act and this must be considered in the analysis. Figure 8.3 taken from reference 153 shows the form which the stress–strain curves of a material may have at differing temperatures, and it is seen that in general, the elastic modulus, limit of proportionality, the secant and tangent modulus after yielding are all functions of temperature. Since the thermal stresses are dependent on the secant modulus, E_s, and temperature, where E_s may itself be dependent on temperature and stress level, an exact inelastic analysis is obviously tedious. Such an analysis is outlined later.

Creep also has an important effect on the behaviour of a hot structure particularly for designs having long exposure times to high stress and temperature. Such effects will not be considered further in this book but they should be considered in practice. For further information on these topics see papers by De Veubeke[26], Hoff[70, 71] and Swanson *et al.*[158]

Thermal fatigue may also occur in structures which are subjected in service to many abrupt changes of temperature causing cyclic thermal stresses, e.g. such conditions occur in the blading of gas turbine engines during starting and shut-down.[20] Repeated cycles of rapid heating or rapid cooling progressively reduce the capacity of the material for plastic deformation (when the temperature gradients are sufficiently large to produce plastic flow) and eventually thermal fatigue may occur. Inelastic analyses may predict the degree of plastic deformation to be experienced in practice, but the main problem is to know the allowable stress for a given life—to this end test data must be acquired. For a most comprehensive discussion on allowable stresses at elevated temperatures, Gatewood's book[48] should be consulted.

8.2. Optimum Design of Structures to Include Thermal Stress

8.2.1. Introduction

Structural design for minimum weight is an aim which, particularly in aeronautics, may have considerable bearing on the economics of operation of the structure. When applied " external " loads occur simultaneously with thermal loads, the problem of optimum design is quite formidable. Not only must the correct choice of structural configuration be made (including the use, or not, of insulation and/or cooling techniques) but the many combinations of applied load and temperature distribution which can occur in practice must all be considered. Obviously no general conclusions can be drawn about optimum design procedures but there are several analyses in the literature dealing with particular structural forms which can be discussed.

8.2.2. Optimum Thickness of a Plate Subjected to Normal Pressure and Temperature Variation through the Thickness

This example, from Przmieniecki's paper,[142] deals specifically with aircraft transparencies for which the design criterion is based on the maximum tensile stress since the materials are brittle. The thermal stress analysis follows that of Sections 2.6 and 4.4 and it is deduced that because of the magnitude of the stresses due to T_m the method of panel support must ensure that such stresses are small by permitting unrestrained thermal expansion in the plane of the panel. Thus, thermal stresses due to ΔT and T^* only need to be considered.

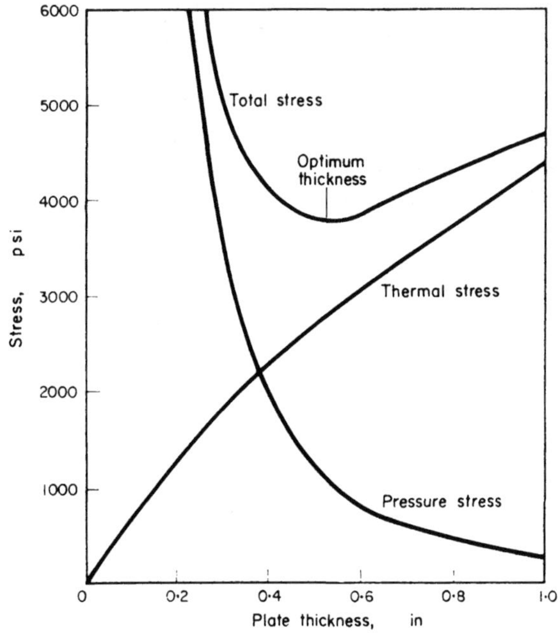

FIG. 8.4. Variation of thermal and pressure stresses in a simply-supported rectangular plate—Taken from reference 142. (*Roy. Aero. Soc.*)

The pressure stresses are maximum when the pressure produces tensile stresses on the external face of the panel and it can be shown that the worst combined loading occurs when the thermal stresses due to ΔT and T^* are also tensile on the external face. This would occur on an aircraft when it is *decelerating* from high speed flight and the magnitude of the maximum tensile thermal stress then increases markedly with increase in plate thickness. Figure 8.4 illustrates a typical set of results and shows how the pressure stresses decrease and thermal stresses increase with increasing plate thickness. The optimum thickness for minimum stress is easily determined, but if a slightly higher stress is acceptable for the given environment and required life (with due allowance for creep, thermal fatigue, etc.) a smaller thickness than the " optimum " could be chosen.

8.2.3. Optimum Design of a Multicell Box Subjected to a Given Bending Moment and Temperature Distribution

In the investigation reported in reference 80 attention is concentrated on the aeronautical application in that the depth of the multicell box (viz. the wing) is pre-fixed by aerodynamic considerations and the temperature distribution considered approximates that likely to be experienced during kinetic heating. The design criterion for the box is taken to be buckling stability of the box skin, of

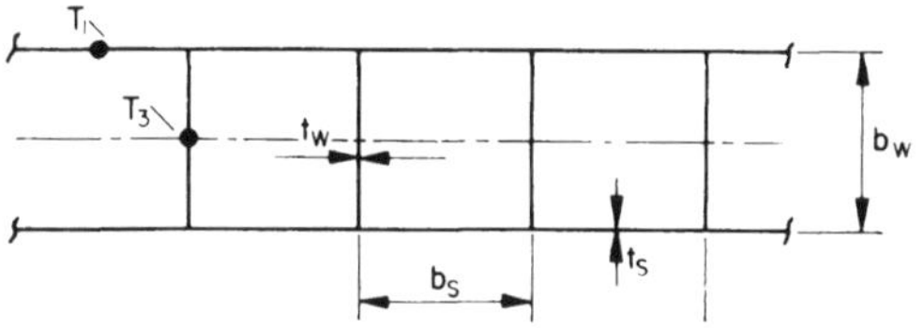

FIG. 8.5. Idealized multiweb beam.

thickness t_s, under combined compressive bending and thermal stresses. The box configuration is shown in Fig. 8.5.

The bending stress in the skin may be written as

$$\sigma_B = m/b_w t_s \left(1 + \tfrac{1}{6}\, r_b r_t\right) \tag{8.2}$$

where m is the bending moment per unit chordwise length (z), and r_b, r_t are the ratios b_w/b_s, t_w/t_s respectively; (see also Figs. 5.2, 5.3)

The thermal stress may be determined from the formulae of Section 5.3 with the assumption that,

(a) there is no temperature variation across the skin, i.e. $T_2 = T_1$,

(b) the material properties are homogeneous and independent of temperature.

Therefore eqn. (5.20) yields

$$\sigma_T = \tfrac{2}{3} E\alpha\, (T_1 - T_3)\, [r_b r_t/(1 + \tfrac{1}{2} r_b r_t)] \tag{8.3}$$

for the compressive thermal stress in the skin.

The buckling criterion requires that,

$$\sigma_B + \sigma_T \ngtr \sigma_{cr} \tag{8.4}$$

where the critical stress is given by

$$\sigma_{cr} = K_1 E_s\, (t_s/b_s)^2 \tag{8.5}$$

K_1 is a function of the parameters r_b, r_t; and E_s is the secant modulus ($= \sigma/e$).

The procedure outlined in reference 80 is to assume values of r_b, r_t and t_s/b_w, and from eqn. (8.4) ($\sigma_B = \sigma_{cr} - \sigma_T$) obtain maximum values of m/b_w^2 as a function of the product $r_b r_t$, r_b and t_s/b_w. Typical curves are shown in Fig. 8.6 for an aluminium alloy (D.T.D.687) structure with a given value of t_s/b_w and the product $E\alpha(T_1 - T_3)$. From Fig. 8.6 a relationship between the " most efficient " m/b_w^2 and the product $r_b r_t$ may be plotted (e.g. Fig. 8.7). Since it can be shown that the weight per unit chordwise length is proportional to $(1 + \frac{1}{2} r_b r_t) t_s/b_w$, then by keeping t_s/b_w and the product $r_b r_t$ constant, the weight is kept constant and the curves in Fig. 8.7 show the relation between the " most efficient " m/b_w^2 and, effectively, the weight, when thermal stresses are both present and absent. The optimum structure for the assumed values of t_s/b_w is that one for which the " most efficient " m/b_w^2 equals the given applied value. If t_s/b_w is not initially specified (e.g. by overall stiffness considerations) the above procedure must be repeated using various values of t_s/b_w until the lightest structure for a given m/b_w^2 is obtained. From Fig. 8.7 it can be deduced that as thermal effects are included in the structure the optimum box geometry, for a given t_s/b_w and m/b_w^2, requires thicker webs more closely spaced.

It should of course be realized that the temperature distribution in the structure is itself a function of the box geometry. This effect is not included in the present analysis but by assuming t_s and b_w to be initially fixed, the two main geometrical parameters governing the most critical temperature distribution during kinetic heating are kept constant.

Another analysis on similar lines is given by McWithey[114].

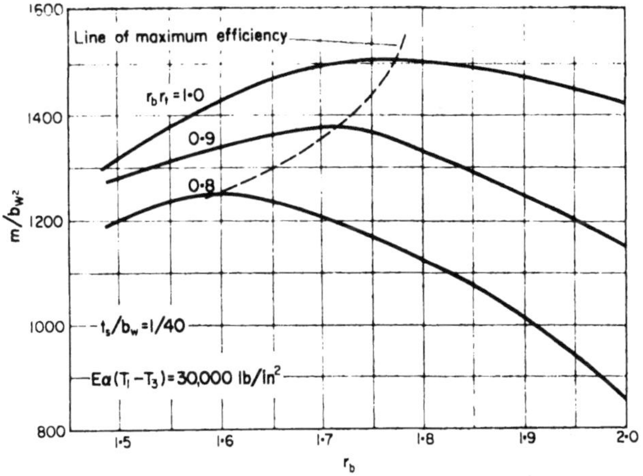

FIG. 8.6. Variation of bending moment parameter with beam configuration.

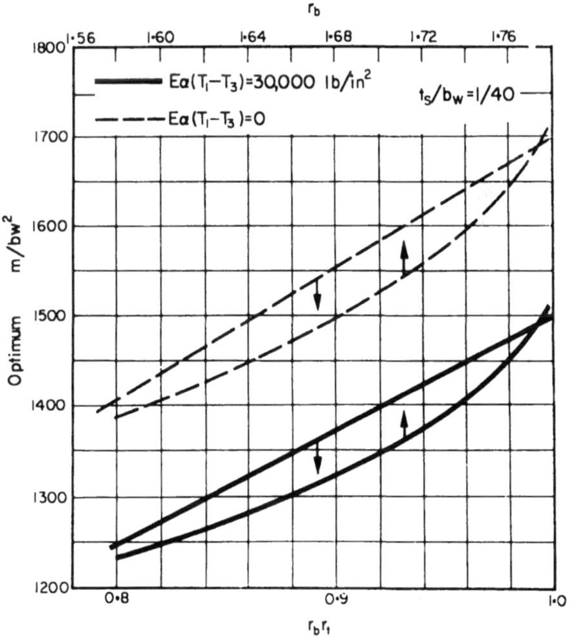

FIG. 8.7. Variation of bending moment parameter with beam weight parameter for different configurations.

8.3. The Alleviation of Thermal Stress

8.3.1. Introduction

In the efficient design of structures to withstand thermal as well as applied loads, consideration must be given to all forms of alleviation of the thermal effects and the weight, complexity, and cost of various alternative schemes must be carefully assessed.

The input of heat into the structure depends primarily on the type of problem and the given environment but it is also influenced by the ingenuity of the designer in that radiation, conduction and convection effects can be used to advantage. Such procedures for the aeronautical application are discussed in references 32, 86, 92, 99 and 169. Obviously, the use of heat-resistant materials and coatings of high emissivity and low thermal conductivity lowers the heat input; as does the use of sandwich forms of structure, such as honeycomb, which make use of the very low conductivity of air.

Assuming a given heat input into the structure the magnitude of the thermal stresses will depend, also, on the relative stiffness of component members and their materials. By careful structural design the thermal stresses can also be reduced.

It is seen therefore that there are two distinct methods by which thermal stresses are alleviated,

(*a*) Reduction of the heat input by thermal insulation and/or cooling, etc.

(*b*) Good structural design, or as Van Der Neut describes it[(169)]— " elastic insulation ".

8.3.2. Thermal Insulation and Cooling

From equations developed in the Appendix it is seen that the efficiency of an insulator is governed mainly by the smallness of its thermal diffusivity (κ). For designs where weight has to be minimized it is more relevant to assess efficiency on the basis of diffusivity and density.

The effect of external insulation on a structure is that a high resistance is provided to the heat flow from an external source. Thus for forced convection heating the external surface of the insulator experiences a rapid rise in temperature which results in a smaller convective heat flux to the surface and, possibly, a larger radiation cooling from the surface. By this means the primary structure experiences an appreciable time lag before its temperature has risen significantly; but, eventually the entire structure must reach a uniform equilibrium temperature, in the absence of any cooling and, thus, a steady state condition of zero thermal stresses.

Should high temperatures be unacceptable on the inner surfaces of the structure some form of forced cooling there is required. It must be appreciated that the most significant result is, then, to establish a steady state temperature gradient through the structure

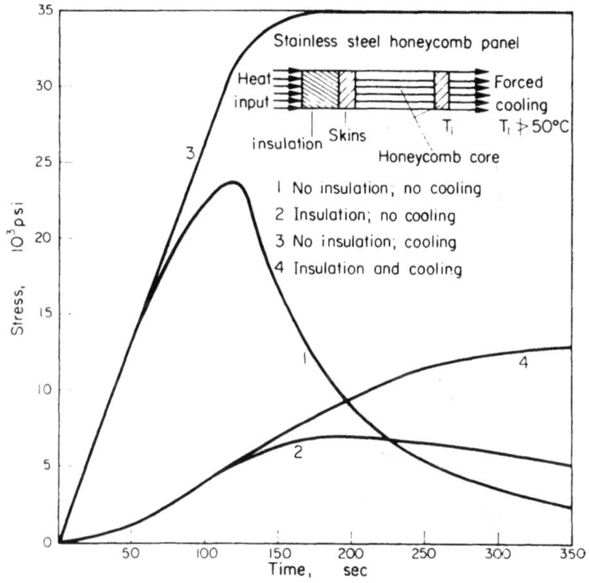

FIG. 8.8. Effects of insulation and cooling on the thermal stresses in a sandwich panel—Taken from reference 92. (*Roy. Aero. Soc.*)

which from the thermal stress point of view may be worse than having an uninsulated and uncooled structure.

Kitchenside[92] presents the results of an analysis into the effects of combinations of external insulation and internal cooling on the temperature gradients and thermal stresses in a stainless steel honeycomb sandwich panel. It is assumed that this is accelerated at 40,000 ft altitude from a Mach No. of 1·5 to 3·0 in 2 min after which the speed remains constant. The maximum adiabatic wall temperature (see eqn. (A. 27)) is approximately 290°C. Figure 8.8 summarizes the results for the thermal stresses.

The beneficial effects of insulation alone are seen from curves 1 and 2 but the use of cooling alone (curve 3) is seen to be disadvantageous, since a stable condition of very high thermal stress is established. Even when insulation and cooling are both used (curve 4) the steady state thermal stresses are twice the peak stresses found for insulation alone. It is assumed that sufficient forced cooling is present to prevent the inner face temperature exceeding 50°C; for cases 3 and 4 the respective values of cooling are, approximately, three-fifths and one-fifth of a kilowatt per square foot. These values and the corresponding cooling system weights, in conjunction with the thickness and type of insulator used, are needed if an attempt is made to optimize the combined insulator–cooling system weight.

Optimization studies for a particular structure using insulation and cooling for long flight times are contained in a paper by Dukes[32] and for short flight times, using insulation only, in the paper by Dobbins[27].

8.3.3. Elastic Insulation

As defined in reference 169, " elastic insulation " is the provision of design features which prevent the occurrence of thermal stresses even though thermal gradients are present.

The thermal gradients can be minimized by the design of joints which offer least resistance to heat flow, and by provision of surfaces having good radiative heat transfer characteristics (see sections, Appendix A3, 5.2V). The provision of secondary heat paths may help to reduce temperature gradients in the adjacent primary structure, e.g. by attaching a secondary web of high conductivity material to the primary web of low conductivity material in a wing structure.

Reduction in restraint between structural components at differing temperatures can be effected in several ways. Sandwich honeycomb construction can have adequate shear strength and stiffness whilst having low extensional stiffness. In a multiweb box design full depth solid webs can be replaced by flanges and post stiffeners, or by a Warren-truss type structure. A corrugated web having low extensional stiffness can be used, which is attached to the skin by discrete cleats. The use of lightening holes in a web lowers the extensional stiffness and hence the restraint and, because of the reduced heat capacity of the web, a more rapid temperature response is experienced by the web resulting in a lower temperature gradient.

For a thin wing with high thermal stresses in the leading and trailing edge structures a straightforward method of removing such stresses is to support these structures from the main structure of the wing by, either, ribs having large flexibility normal to their plane or the use of " piano hinge " type attachments which offer infinite flexibility parallel to the hinge.

The moral for the designer is, therefore: to alleviate thermal stresses—prevent them occurring! Parkes(134) makes observations on these lines and shows how thermal stresses are prevented in bridge and piston engine design. In the former case overall changes in length of the bridge are accommodated freely, by simply supporting one end of the bridge on rollers. In the latter case, the use of a piston ring provides the necessary flexibility to accommodate differences in radial expansion of the piston and cylinder. It should be noted that statically determinate structures do not suffer from thermal stress, and the use of a truss-web, or corrugated web, structure as described above is an attempt to achieve this condition.

For structures which have to be made redundant, or statically indeterminate, a further device is available to minimize the thermal stresses; that is to take advantage of the different thermal expansion properties of different materials when designing the structure.

If the simple analysis of Section 1.1 is considered, it is seen, eqn. (1.4), that σ_1 is zero when $\alpha_1 T_1 L_1 = \alpha_2 T_2 L_2$. Therefore if this equation is satisfied for the service condition of the structure, no stresses result. If, further, it is assumed that $A_2 \gg A_1$ and T_1 is zero then $\sigma_2 = 0$ and σ_1 is given by

$$\sigma_1 = E_1 \alpha_2 T_2 L_2 / L_1 \tag{8.6}$$

To minimize σ_1 the product $E_1 \alpha_2$ must be as small as possible. Similar results can be obtained from the I-beam analysis of Section 5.3, where area A_2 corresponds to the flange area, which in wing design is usually large compared with the web area (A_1) in which the largest thermal stresses (σ_1) usually occur. Parkes(134) examines combinations of six materials for the product $E_1 \alpha_2$ and the results are reproduced in Table 8.1.

It is seen that there are seven combinations of materials for which the product $E_1 \alpha_2$ is less than 100 lb/in^2°C with aluminium (1) and molybdenum (2) forming the best combination. Of the more conventional aircraft materials the Al-steel combination gives a value for $E_1\alpha_2$ of 110 compared with values of 220 and 330 for all aluminium and all-steel structures respectively.

This effect is examined more closely in an example in reference 92 and Fig. 8.9 indicates the results obtained. A multiweb box structure representative of the wing of a supersonic fighter aircraft is considered and Fig. 8.9 shows the thermal stresses calculated at the web centre and skin panel centre (midway between webs) following an instantaneous change of speed. Three combinations of materials are chosen with the proportions of the box structure adjusted in each case to give an (approximately) optimum design for a given external loading, box depth and torsional stiffness.

TABLE 8.1 VALUES OF THE PRODUCT $E_1\alpha_2$ FOR VARIOUS MATERIAL COMBINATIONS(134)

Material 1	E_1†	α_1†	$E_1\alpha_2$					
Aluminium	10	22	220	120	50	110	80	60
Beryllium	44	12	968	528	220	484	352	264
Molybdenum	46	5	1012	552	230	506	368	276
Steel	30	11	660	360	150	330	240	180
Titanium	16	8	352	192	80	176	128	96
Zirconium	13	6	286	156	65	143	104	78
	Material 2		Al	Be	Mo	Steel	Ti	Zi

The high, peak thermal stresses in the all-steel structure gradually decrease as the entire structure reaches a uniform equilibrium temperature condition. The same is true of the all-dural structure but the peak stresses are much smaller due to the higher thermal conductivity and lower value of $E\alpha$. For a steel skin-dural web combination the temperature gradients are smaller than in the all-steel case, due to the higher web thermal conductivity, so that in the transient heating phase only small thermal stresses are caused. In the equilibrium temperature condition the higher coefficient of expansion of dural causes a stable condition of high compressive stress in the web. On subsequent cooling of the skins, even higher web compressive stresses result with obvious disadvantages as regards web stability. Thus a composite structure of the type

† Units: E: 10^6lb/in^2; α: 10^{-6}/°C. Values at 0°C.

considered is advantageous if the temperatures are high enough to warrant the use of steel skins and if the flight times are short enough that unacceptably high web temperatures or compressive thermal stresses do not result. An obvious application might be for a guided missile design.

In the above discussion it is assumed that the thermal stresses are elastic, but it can be shown[48] that one possible way to alleviate

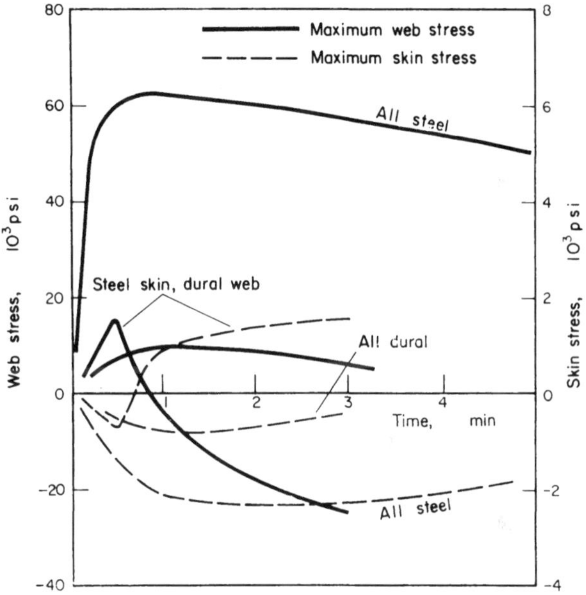

FIG. 8.9. Thermal stresses in multiweb wings for various material combinations—Taken from reference 92. (*Roy. Aero. Soc.*)

thermal stress is to let the inelastic portion of the stress–strain curve provide the necessary deflexion to absorb the thermal expansion. This concept is now examined.

8.4. Inelastic Thermal Stresses

Inelastic thermal stresses may be defined as those which are determined for a variable E distribution, where the variation in E may be due to temperature and/or the limit of proportionality being exceeded. The former effect is considered in Section 8.1 and possible modifications to elastic thermal stress analysis proposed.

Inelastic thermal stresses can occur in partially or fully restrained homogeneous structures at uniform temperature where both of the above effects may be present. Analyses for such problems are relatively straightforward since all elements of the structure have the same stress–strain curve for each value of temperature. In general this does not apply for non-uniform temperature distributions.

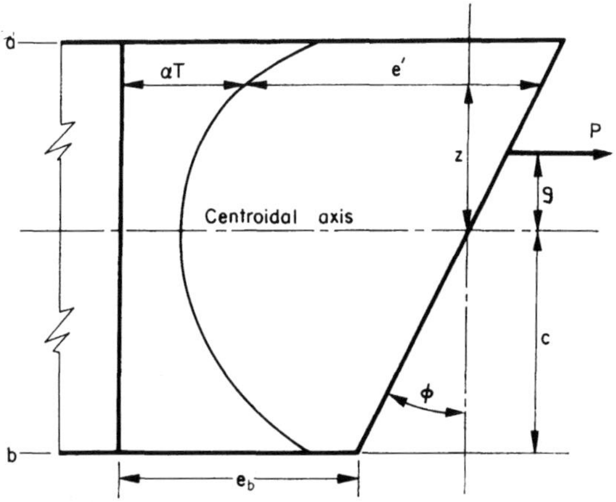

Fig. 8.10. Strain diagram for a beam subjected to end load and a temperature distribution—Taken from reference 153. (*Soc. Auto. Engrs.*)

Several authors have applied inelastic analysis methods to stress systems produced by thermal effects combined with external loading. Usually, the conventional simplifying assumptions of beam theory are observed and the basis of strain analysis is adopted.[47–51, 107, 131, 132, 134, 135, 153, 158]

Among the configurations which have been examined are two- and three-bar systems, columns, plates, ring frames and I-section beams. To illustrate the analytical approach the derivation due to Sprague and Huang[153] is described, based on the beam configuration of Fig. 8.10 in which the temperature varies only in the z direction. Creep effects are not considered here; see reference 158 for a similar analysis in which such effects are included.

As the cross-section of the beam strains it rotates and remains plane. Therefore the total strain, consisting of strain due to stress and expansion, is

$$e_b + \phi(z + c) = e' + \alpha T \tag{8.7}$$

where ϕ is a small angle. The normal stress at any point is given by

$$\sigma = E_s e' = E_s \left[e_b + \phi(z + c) - \alpha T\right] \tag{8.8}$$

and the overall equilibrium conditions give,

$$P = \int_b^a \sigma \mathrm{d}A \tag{8.9}$$

$$Pg = \int_b^a \sigma z \mathrm{d}A \text{ or } P(g + c) = \int_b^a \sigma(z + c)\, \mathrm{d}A \tag{8.10}$$

Therefore using eqn. (8.8)

$$P = e_b \int_b^a E_s \mathrm{d}A + \phi \int_b^a E_s(z + c)\mathrm{d}A - \int_b^a E_s \alpha T \mathrm{d}A \tag{8.11}$$

$$Pg = e_b \int_b^a E_s z \mathrm{d}A + \phi \int_b^a E_s(z + c)z\mathrm{d}A - \int_b^a E_s \alpha T z \mathrm{d}A \tag{8.12}$$

from which, if $\int_b^a E_s z \mathrm{d}A = 0$ defines the position of the neutral axis of the effective cross-section, then,

$$\int_b^a E_s(z + c)\mathrm{d}A = c \int_b^a E_s \mathrm{d}A \tag{8.13}$$

and,

$$P + \int_b^a E_s \alpha T \mathrm{d}A = e_b \int_b^a E_s \mathrm{d}A + \phi c \int_b^a E_s \mathrm{d}A \tag{8.14}$$

$$Pg + \int_b^a E_s \alpha T z \mathrm{d}A = \phi \int_b^a E_s z^2 \mathrm{d}A \tag{8.15}$$

Solution of these two equations for ϕ and e_b yields

$$\phi = \frac{Pg}{E_b I_{kk}} + \int_b^a \frac{k_e k_p \alpha T z \mathrm{d}A}{I_{kk}} \tag{8.16}$$

$$e_b = \frac{P}{E_b}\left(\frac{1}{A_{kk}} - \frac{gc}{I_{kk}}\right) + \int_b^a \frac{k_e k_p \alpha T \mathrm{d}A}{A_{kk}} - c\int_b^a \frac{k_e k_p \alpha T z \mathrm{d}A}{I_{kk}} \quad (8.17)$$

where E_b is the modulus of elasticity at the point b

k_e = modulus of elasticity at any point$/E_b = E/E_b$

k_p = secant modulus at any point/modulus of elasticity at any point

$= E_s/E$.

I_{kk} is the second moment of area of effective cross-section about the neutral axis, and A_{kk} is the total effective area where

$$\left.\begin{aligned} I_{kk} &= \int_b^a \frac{E_s}{E_b} z^2 \mathrm{d}A = \int_b^a k_e k_p z^2 \mathrm{d}A \\ A_{kk} &= \int_b^a \frac{E_s}{E_b} \mathrm{d}A = \int_b^a k_e k_p \mathrm{d}A \end{aligned}\right\} \quad (8.18)$$

It should be noted that if there is an additional bending moment, M, on the section, then eqn. (8.10) is modified by adding M to the left-hand side.

An iterative procedure is outlined in reference 153 for solution of a given problem, where it should be noted that both ϕ and e_b are functions of the elasticity factors k_e and k_p. These factors vary with the temperature and degree of plasticity over the cross-section, and also on the incremental direction of loading at each section (see below). The curves in Fig. 8.3 illustrate stress–strain relationships for one material at various temperatures and stress levels.

If initial values are assigned to ϕ and e_b (e.g. the elastic values) eqn. 8.8 enables the stress distribution to be found for a given temperature distribution $T = T_{(z)}$. The stress–strain relations (e.g. Fig. 8.3) must be known to determine the appropriate E_s distribution for the given strain distribution e'. Overall equilibrium is checked using eqns. (8.9) and (8.10) and any unbalance is adjusted by making corrections to the values of ϕ and e_b. The new values of ϕ and e_b can be found from eqns. (8.16) and (8.17) with the parameters c (and hence g), k_e, k_p, A_{kk}, I_{kk} determined from established data from the previous set of calculations. Obviously if elastic property data are used to determine the initial assumed values of ϕ and e_b, then the first iteration gives the correct answer if the solution is entirely elastic.

If during a loading and/or heating cycle any section, dA, experiences a reduction in e' between one time interval and the next this is termed " unloading " since it must be accompanied by a reduction in the stress σ. The behaviour of that section during this

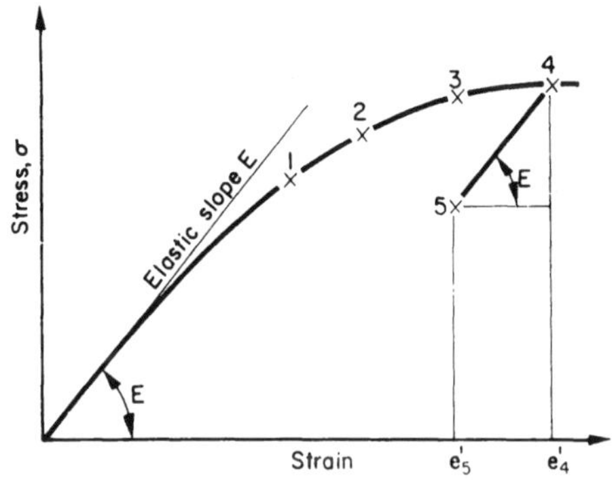

Fig. 8.11. Elastic unloading of a structural element which has yielded.

process is elastic and the new stress (σ_5) is related to the original value (σ_4)—see Fig. 8.11—by

$$\left.\begin{aligned} \sigma_5 &= \sigma_4 - E(e_4' - e_5') \\ &= Ee_5' - (E - E_{s4})e_4' \end{aligned}\right\} \qquad (8.19)$$

Obviously, the iterative method of analysis can be adapted for high-speed digital computation over the full range of stress–strain relations up to the failing load and for cyclic loading or heating.

The shape of the elastic stress distribution over the cross-section of a rectangular plate for a given tensile loading and/or heating is indicated in Fig. 8.12 by curves B and/or A. As the tensile load is increased the most highly stressed fibres deform plastically due to inelasticity and there is a corresponding change in the shape of the stress distribution. It is seen that the initial stress distribution due to the tent-like temperature distribution (curve A) is gradually smoothed out so that at the higher tensile loads the stress distribution is almost uniform. This suggests that the initial non-uniform thermal stress distribution does not significantly affect the static tensile load for a ductile material at failure.[48,153] Thus it may be concluded

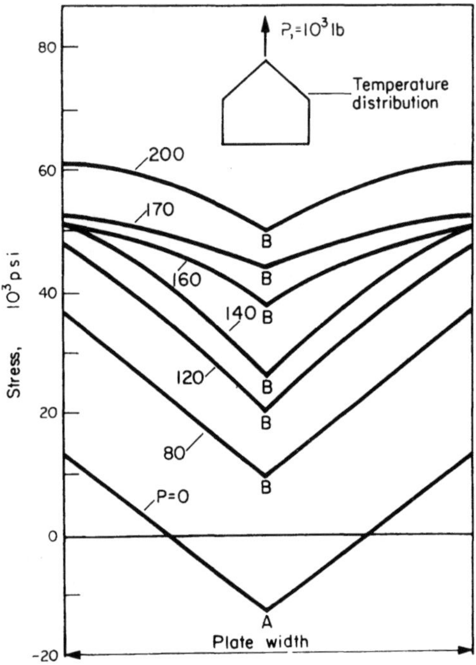

FIG. 8.12. Longitudinal stress distribution in a plate showing progressive yielding with increasing tensile load—Taken from reference 153. (*Soc. Auto. Engrs.*)

that the failing loads for structures of ductile materials under non-uniform temperature distributions can be determined from a knowledge of the external loading only, provided that the correct temperature-dependent properties of the material are used.

In this way inelasticity absorbs the effects of thermal stress. It should be realized however that in the subsequent unheated and unloaded condition the material sustains a residual stress distribution. The magnitude of this distribution is essentially the same as the initial elastic thermal stress distribution (but of opposite sign) if the previous subsequent external loading is such that this initial distribution is completely smoothed out and diminished.

To determine the residual stresses in a structure an incremental method of solution is used based on the inelastic analysis given earlier, and eqn. (8.19). Working from the condition of a heated and loaded structure subjected to plastic deformations, the principle of superposition is applied to remove sequentially the external loading and the temperature distribution.[153] The load is removed by applying an equal and opposite force for which the corresponding stress distribution is determined; the temperature is removed similarly by determining the stress distribution corresponding to a set of equal and opposite thermal strains. The residual stress pattern at any stage is found by adding these last two stress distributions algebraically to the original plastically deformed system. For the many possible unloading and cooling sequences an incremental solution using the above procedure is necessary, with care being taken for each increment to unload and cool in the correct sequence, since this affects the final residual stress pattern.

Therefore it is seen that the non-linear part of the stress–strain curve can accommodate the thermal strains if design for some permanent set is permissible. For a single cycle of thermal loading this philosophy is now accepted for ductile materials, but caution must be exercised if the thermal loading cycle is to be repeated many times, as then other considerations arise, of the nature of thermal fatigue for example.

8.5. Cyclic Thermal Loading : Incremental Collapse

When a structural element originally in a state of equilibrium, is loaded, by either mechanically or thermally induced stresses, a stress–strain state is established in the material of the element which satisfies the new equilibrium condition. If as a result yielding occurs at some sections in the element, then a different equilibrium condition from the original is obtained when the external loading and/or temperature distribution is removed. The difference is a measure of the residual stress pattern which may cause dimensional changes in the element as described in the previous section.

If the changes in dimension occur on the first cycle only, but thereafter the stresses are always elastic in subsequent cycles, the behaviour is termed " simple shakedown ". If however there are further changes in subsequent cycles, but these gradually decrease in magnitude to zero, " gradual shakedown " is said to occur.

When there is both positive and negative plastic flow in successive cycles, with zero net value, " alternate plasticity " occurs; but if the net value of plastic flow following each cycle is not zero, and remains constant or increases with increasing number of cycles then " incremental collapse " is said to occur. The element continues to change its dimensions and shape progressively until the deformation becomes unacceptable, or failure occurs.

This phenomenon of incremental collapse is a mechanical effect caused by the establishment of a *ratchet mechanism* which only permits the growth of strain in one direction. This mechanism has been considered for various engineering applications, e.g. aircraft(131, 132, 134, 135, 153, 165) and pressure vessels(116) where thermal effects are present in each case. Other relevant papers are those of Neal(123, 124).

In references 131 and 132 a particular form of incremental collapse is considered by Parkes such as might occur in an aircraft wing. The wing is subjected to a steady stress system due to an applied spanwise bending moment on which is superimposed thermal stress cycles as might be experienced due to successive accelerated and decelerated supersonic flight. The incremental collapse mechanism is found to be one of steadily increasing spanwise curvature and length.

The multicell wing, see Fig. 5.2, is represented by a typical, rectangular box beam and has an initial, elastic, bending stress distribution. Starting from a uniform (zero), reference temperature level the skin first acquires a temperature T, following which the web temperature rises (uniformly) to a value γT. In succession, the skin and web temperatures then drop to zero to complete the temperature cycle which may be illustrated symbolically by,

$$(\text{skin } 0, \text{ web } 0) \rightarrow (\text{skin } T, \text{ web } 0) \rightarrow (\text{skin } T, \text{ web } \gamma T) \rightarrow$$
$$(\text{skin } 0, \text{ web } \gamma T) \rightarrow (\text{skin } 0, \text{ web } 0).$$

Since the combined mechanical and thermal stress distribution is antisymmetric, yielding occurs asymmetrically when the thermal stresses are sufficiently high. The collapse rates per cycle as given in reference 132 are :—

compression (upper) skin strain decreases by $(1 - \beta)\alpha T$,
tension (lower) skin strain increases by

$$(1 + \gamma - \beta)\, \alpha T - (2\sigma_y - \sigma_1 + \sigma_F)/E,$$

and, hence, the mean longitudinal strain changes by

$$\tfrac{1}{2}\gamma\alpha T - \tfrac{1}{2}(2\sigma_y - \sigma_1 + \sigma_F)/E.$$

In these expressions βT is the temperature at which the yield stress of the material drops from σ_y to σ_F. Then, σ_F is the upper skin stress and σ_1 is the corresponding lower skin stress at the condition (skin βT, web 0) where $\gamma \leqslant \beta \leqslant 1$.

Parkes assumes that there are no temperature gradients in either the skins or webs and that the material has a definite yield stress with perfect plasticity. For a more realistic stress–strain curve and temperature distribution a similar method to that outlined in Section 8.4 is necessary which requires an iterative procedure for its solution.

In order to assess the significance of the above assumptions reference 165 contains details of an experimental investigation and of an associated theoretical development which permits the inclusion of more realistic temperature distributions and stress–strain data, and yet does not require an iterative technique.

The stress–strain curves for each temperature are assumed to consist of three linear sections as shown in Fig. 8.13, taken from reference 165. A closed series of instantaneous temperature distributions define the thermal cycle and associated with each are corresponding values of the external loading quantities. Between two adjacent points in time in the above series (a " step ") the load quantities and temperatures are assumed to vary linearly with a common parameter λ (which may often correspond to time).

The beam cross-section is divided into segments which are assumed to experience uniform temperature and direct strain. The temperatures correspond to the assumed distributions and the strains always satisfy the condition of linearity through the beam depth. Each " step " may consist of a series of " sub-steps " during which there is a linear variation of stress with strain and temperature for any segment.

Because of the various linearities imposed in the analysis, the relationships between stress and strain for all strains and temperatures can be more easily stored as information in the digital computer used; and if the overall strain is defined by, say, a mean strain e and curvature ϕ then both these quantities are found to vary linearly with λ. The " steps " are initially prefixed but the " sub-steps " are bounded by situations where a beam segment passes from one linear section of the appropriate stress–strain curve to another, in Fig. 8.13. Such changes are initiated by one of the segments

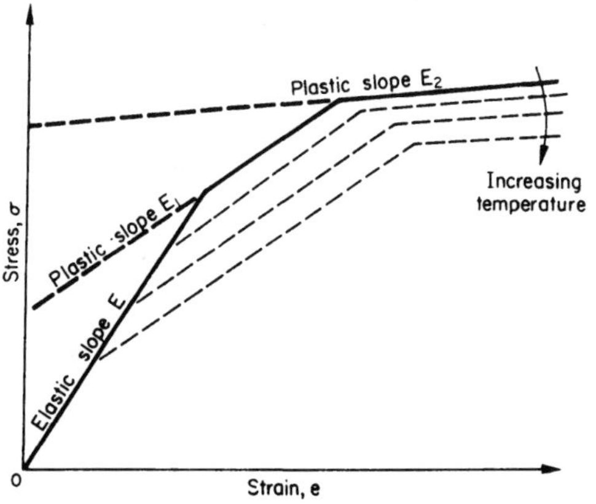

FIG. 8.13. Definition of assumed stress–strain relationships—Taken from reference 165.

returning from plastic to elastic behaviour as shown in Fig. 8.11 at point 4. This latter effect may also occur at the beginning of a " step " where changes are made in the temperature rates of change. Thus e and ϕ vary through each temperature change to form a curve made of linear sections.

Obviously the constancy of the slopes of the three linear sections of Fig. 8.13 cannot be justified for a wide temperature range, but since it produces a less tedious method of solution it has some merit. However, a more serious limitation to any method of solution can be seen even when data of the type shown in Fig. 8.3 are used. It has already been emphasized in Section 8.1 that such data are strongly dependent on strain–rate and temperature-rate, so that the choice of such, more realistic, data (as in Fig. 8.3) is difficult.

For any beam segment, although the temperature rate may be known, the strain rate is to be determined and is not initially known. Also data of the type shown in Fig. 8.3 are usually obtained by keeping temperature constant and varying stress and strain at a particular rate. The applicability of such data to situations where all three

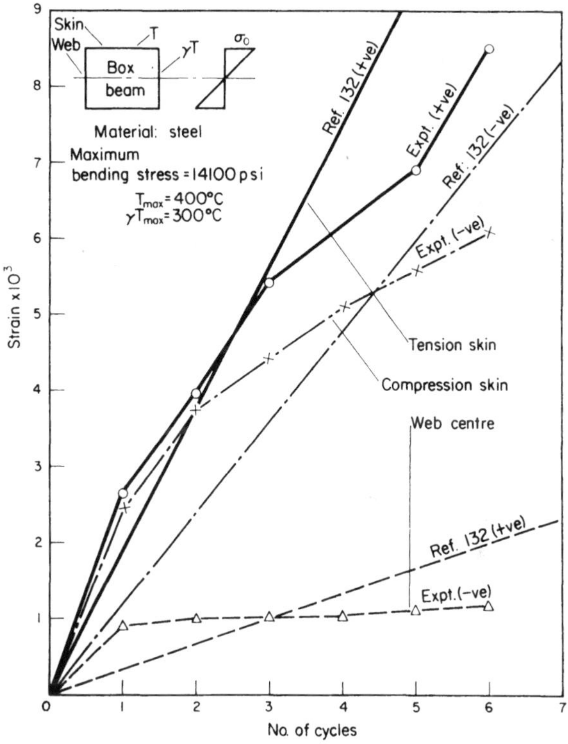

FIG. 8.14. Total residual strains in a box beam after successive thermal cycles—Taken from reference 165.

parameters are changing progressively at initially unknown rates (for stress and strain) is obviously suspect. Much more extensive research is needed in this field.

By making some reasonable assumptions the analysis of Parkes is applied in reference 165 to attempt a correlation with an experimental investigation. Figure 8.14 shows the results obtained for the collapse rates per cycle and Fig. 8.15 is a photograph of the " collapsed " specimen showing the curvature in the beam.

In reference 135 a very simple redundant structure (a two-bar system) is considered, which is subjected to temperature cycling, to determine the influence of the yield stress–temperature relation

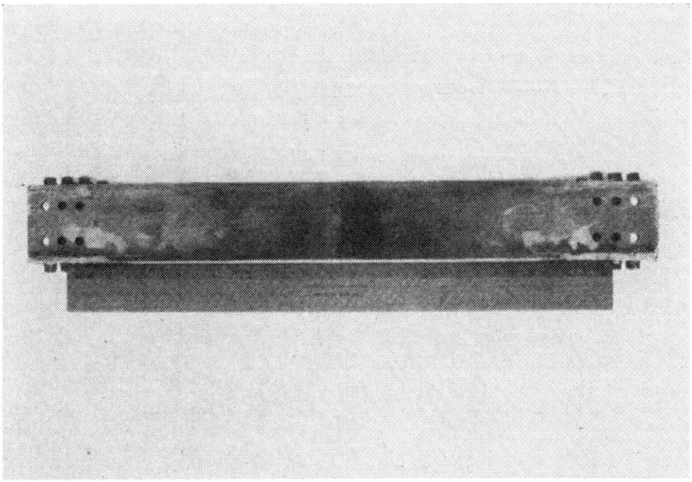

FIG. 8.15. A " collapsed " specimen showing curvature—Taken from reference 165.

on the behaviour, with the range and periodic time of the temperature cycle as additional variables. It is found that improving the strength of the material at elevated temperatures may have the undesirable effect of hastening incremental collapse of the structure. Also, the most rapid incremental collapse is not necessarily associated with the maximum values for the range or periodic time of the temperature cycle. It is also found that there can be a discontinuity in behaviour between a structure made from a material with a small negative strength/temperature gradient and one with a small positive value.

Simple two-bar systems are considered also in references 107, 116 and 134. Reference 134 shows that if one of the two bars is much larger than the other, the larger member always remains elastic and incremental collapse, which requires plasticity in both members, cannot occur. However, shakedown or alternate plasticity may occur due to inelasticity in the smaller member.

Formulae for incremental growth in two- or three-bar systems are given in several papers(116, 134, 135) and will not be repeated here in detail. For the three-bar system shown in Fig. 8.16 the growth strain per cycle, once a steady cycle has been established, is given

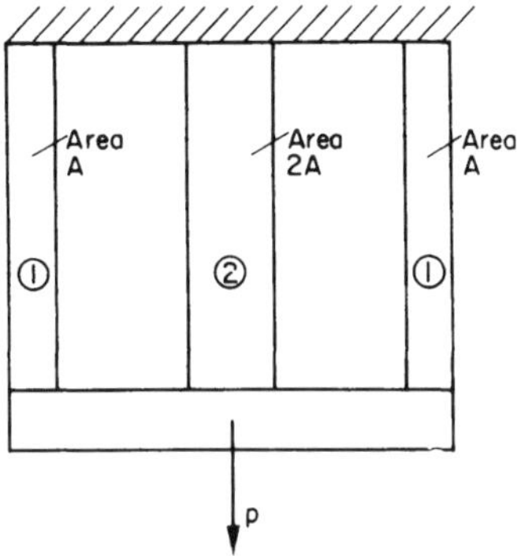

FIG. 8.16. A simple three-bar structure.

below. The thermal cycle consists of successive heating and cooling of the outer bars (1) to and from a temperature, T, above the constant temperature of the inner bar (2). The growth strain per cycle for a material with perfect plasticity is

$$e = \frac{\sigma_y}{E}\,[2m - 4(1 - n)] \qquad (8.20)$$

where σ_y is the yield stress, assumed invariant in the temperature range

$$m = E\alpha T/2\sigma_y$$

$$n = \sigma_m/\sigma_y \; (1 \geq n > 0)$$

$$\sigma_m = \text{tensile stress due to applied load } P.$$

It is seen that when $n = 1$, i.e. the applied load P is sufficient to just cause tensile yielding in all bars, the strain growth per cycle is simply equal to the thermal strain, αT. Even when there is no external loading ($n = 0$) growth occurs when $m > 2$. (Note: e negative indicates zero growth)

Other cases where growth might be predicted on the basis of analyses similar to those described above are;[107]

(1) Pressure vessels subjected to temperature transients in their contained fluid heated either externally or internally (e.g. nuclear reactors).

(2) Temperature cycling of pressure vessels clad with material having a different coefficient of thermal expansion.

(3) Bimetallic joints in pressure piping subjected to temperature cycling (materials with different expansion characteristics).

(4) Temperature cycling of pressure vessels made up of a material consisting of crystal phases differing in thermal expansion characteristics.

(5) Turbine blades and discs subjected to severe temperature gradients and centrifugal loading.

(6) Pressurized piping subjected to high bending or torsional loadings (e.g. bellows).

In connexion with (4) it is reported in reference 107 that rolled uranium rods may exhibit a strain growth per cycle under cyclic temperature only; the growth may be attributed to intergranular stresses set up on heating by the anisotropy of thermal expansion of the crystals. Reference 107 also describes another unusual type of growth in which a long thin-walled tube of austenitic steel was filled, when vertical, with fine ferritic steel shot. On being heated the tube expanded and the shot settled down, whereas on cooling the shot could not be forced up the tube. Thus, the tube yielded, causing incremental growth of diameter, with successive heating cycles.

APPENDIX

Heat Transfer in Structures

THE basic modes of heat transfer are conduction, convection and radiation of which conduction is the primary mode, with the other two only affecting the boundary conditions. In this appendix all three modes will be considered and procedures outlined for the determination of transient and equilibrium temperature distributions in practical structures.

A.1. Heat Conduction

In a solid body heat is transferred mainly by conduction, since the effects of radiation are negligible except for transparent materials such as glass or quartz. The theory of heat conduction is based on Fourier's law, which for an isotropic body may be written,

$$q = -k\frac{\partial T}{\partial n} \tag{A.1}$$

where n is normal to the surface, or, more specifically in rectangular coordinates,

$$(q_x, q_y, q_z) = -k\left(\frac{\partial T}{\partial x}, \frac{\partial T}{\partial y}, \frac{\partial T}{\partial z}\right) \tag{A.2}$$

where the heat flux per unit area, q, is in the direction of positive normal, or positive coordinate and k, the thermal conductivity, may be a function of temperature. Typical units for k are C.H.U/(hr ft °C).

The flux δQ per unit volume, into a small element of volume in time δt, is given by

$$\delta Q = -\left(\frac{\partial q_x}{\partial x} + \frac{\partial q_y}{\partial y} + \frac{\partial q_z}{\partial z}\right)\delta t \tag{A.3}$$

From the Second Law of Thermodynamics the relation

$$\delta Q = T'\mathrm{d}\eta \tag{A.4}$$

is obtained, where η is the entropy per unit volume and T' is the absolute temperature given by

$$T' = T + T_0 \tag{A.5}$$

when T is the temperature rise from the initial, uniform, temperature T_0, of the stress-free state.

Combining eqns. (A.2), (A.3) and (A.4) yields the result,

$$T'\frac{\partial \eta}{\partial t} = \frac{\partial}{\partial x}\left(k\frac{\partial T}{\partial x}\right) + \frac{\partial}{\partial y}\left(k\frac{\partial T}{\partial y}\right) + \frac{\partial}{\partial z}\left(k\frac{\partial T}{\partial z}\right) \tag{A.6}$$

and in general, the thermal conduction problem is linked through η to the state of strain or stress. This is shown in more detail in Section 1.6, but when thermoelastic coupling is negligible, as it usually is, eqn. (A.6) becomes simply

$$C\frac{\partial T}{\partial t} = \frac{\partial}{\partial x}\left(k\frac{\partial T}{\partial x}\right) + \frac{\partial}{\partial y}\left(k\frac{\partial T}{\partial y}\right) + \frac{\partial}{\partial z}\left(k\frac{\partial T}{\partial z}\right) \tag{A.7}$$

where C is the heat capacity per unit volume.

If the body also had internal heat generation at the rate H per unit volume per unit time, this term would enter the right-hand side of (A.7), and if k is sensibly constant

$$\frac{\partial T}{\partial t} = \kappa \nabla^2 T + H/C \tag{A.8}$$

where $\kappa(= k/C)$ is known as the thermal diffusivity of the material, with typical units of $\mathrm{ft}^2/\mathrm{hr}$, and $\nabla^2 T$, the Laplacian operator, is written as

$$\nabla^2 T = \frac{\partial^2 T}{\partial x^2} + \frac{\partial^2 T}{\partial y^2} + \frac{\partial^2 T}{\partial z^2} \tag{A.9}$$

For the other coordinate systems $\nabla^2 T$ may be written accordingly[15] using the transformations given on Figs. 1.2 and 1.3.

Equation (A.7) was obtained by considering the heat balance of an element of volume $\mathrm{d}x\mathrm{d}y\mathrm{d}z$, and for the three-dimensional problem is sufficiently general. When k is constant the corresponding one-dimensional result for heat conduction along a bar of variable cross-sectional area $A = A(x)$ is

$$C\frac{\partial T}{\partial t} = k\frac{\partial^2 T}{\partial x^2} + \frac{k}{A}\frac{\partial A}{\partial x}\frac{\partial T}{\partial x} = \frac{k}{A}\frac{\partial}{\partial x}\left(A\frac{\partial T}{\partial x}\right) \qquad (A.10)$$

This result may be obtained directly from (A.7) or by considering the heat balance of a typical element, of length dx and having the shape of a conical frustrum. The following result is obtained when A is constant,

$$C\frac{\partial T}{\partial t} = k\frac{\partial^2 T}{\partial x^2} \qquad (A.11)$$

The solution of the above partial differential equations (A.7, A.8, A.10 or A.11) with the appropriate boundary conditions, is possible by various analytical methods, which for many simple problems are quite straightforward in their application. It may for instance be possible to use the Laplace transformation method which eliminates the time variable by an integration. A boundary value problem then remains for the space variables which on being solved must be followed by an inversion in the complex plane to recover the time variable. This inversion is often quite difficult and the final solution may necessitate numerical integration.

However, in many practical problems of interest, where additional complexity may be present due to temperature dependent material properties or because of non-linear boundary conditions, these analytical methods have very little practical applicability. In such cases the nonlinear heat conduction problem is more easily solved by finite difference techniques. For this reason no discussion of the various analytical methods is given here, but the finite difference formulation is discussed more fully later. The reader interested in a more detailed discussion of all the various methods of solution is referred to the books of Boley and Weiner[15], Carslaw and Jaeger[18], Eckert and Drake[36], Jakob[79] and McAdams[111].

To solve any given problem it is, of course, necessary to specify the boundary conditions on the body surface, and the initial conditions throughout the body. These latter conditions refer to the initial temperature distribution which is usually uniform. The boundary conditions, of which there are five, describe the possible physical processes that may occur on the surface. To avoid repetition these five conditions will be defined in Section A.5 of this Appendix when their use in a finite difference formulation will be described.

A.2. Heat Convection

For heat convection from a fluid to a solid, the heat flux at the solid surface may be written as

$$q = h\Delta T \tag{A.12}$$

where h is termed the heat transfer coefficient and ΔT is a representative temperature difference between the fluid and the solid. Typical units for q would be C.H.U./ft^2 hr.

Unfortunately, h is not a constant quantity and since the fluid temperature may vary significantly across the boundary layer decisions have to be taken on the values to be ascribed to both h and ΔT.

A.2.1. Free Convection

In this case the fluid motion is due to density changes in the fluid occasioned by the heat transfer from solid to fluid and the temperature difference ΔT is taken to be that existing between the fluid outside the temperature boundary layer and the solid body surface. It can also be shown that h depends very strongly upon the temperature difference ΔT; in fact for a vertical surface[36]

$$h \propto \Delta T^{1/4} \tag{A.13}$$

In general, heat transfer by free (or natural) convection is quite small.

A.2.2. Forced Convection

Heat transfer by forced convection to or from a solid surface is due to the motion of the fluid past the surface. Basically two types of flow near to the surface can occur. For laminar flow the fluid moves mainly in the main-stream direction in layers parallel to the surface, whereas turbulent flow exists when the fluid particles have random motions in all directions, including the main stream direction, such that mixing of the fluid layers occurs.

The heat transfer at the surface takes place by molecular conduction in the fluid layers adjacent to the surface whether the flow is laminar or turbulent. For the latter case, however, heat transfer also occurs by fluid transport due to the random motions of the turbulent eddies. Thus if the rate of heat transfer at the body surface is defined by

$$q = -k_F \frac{\partial T_F}{\partial n} \qquad \text{(A.14)}$$

where k_F is the thermal conductivity of the fluid, turbulent motion has the effect of steepening the temperature gradient in the fluid near to the surface, thereby increasing the heat transfer rate.

As mentioned earlier, eqn. (A.12) is used to express the heat transfer rate at the surface, where h, the heat transfer coefficient, is dimensional and varies considerably with conditions in the flow, e.g. the boundary layer state, etc.

The boundary layer concept has proved most useful in formulating analyses of forced convection since two regions in the fluid may be considered independently. In the thin boundary layer near to the surface consideration is given to viscous effects and these together with heat conduction terms are retained in the analysis. Away from the body the effects of viscosity and conduction are negligible since the gradients of temperature and velocity are themselves negligible outside the boundary layer.

It can therefore be shown that the heat transfer coefficient, h, is a function of many parameters which are themselves functions of the flow because of the variations in temperature through the fluid. To determine the dependence of h on these various parameters, e.g. the thermal conductivity, k_F; coefficient of viscosity, μ; density, ρ; etc., dimensional analysis can be used,[48] with consideration of all the parameters that would enter the more detailed fluid flow equations. Let

$$h = G U_\infty^a L^b \mu^f k_F^j \rho^m C_p^n \qquad \text{(A.15)}$$

where U_∞ is the flow velocity, L is a dimension of length, C_p is the specific heat at constant pressure and G is a constant. The above equation is rewritten so that all parameters are expressed in the basic units of heat energy, time, length, mass and temperature. By equating the exponents for each basic unit, the number of unknown exponents can be reduced to two; hence

$$N_u = G.R_e^m P_r^n \quad (A.16)$$

where $N_u = \dfrac{hL}{k_F}$ is the Nusselt number,

$R_e = \dfrac{U_\infty L\rho}{\mu}$ is the Reynolds number and

$P_r = \dfrac{\mu C_p}{k_F}$ is the Prandtl number.

The parameters G, m and n may be determined by theoretical analysis or by experiment. Thus, for incompressible laminar flow over a flat surface of length L in the main stream direction, the local Nusselt number is given by,

$$N_u = \frac{hL}{k_F} = 0{\cdot}332 R_e^{1/2} P_r^{1/3} \quad (A.17)$$

An alternative non-dimensional quantity to N_u is the Stanton number, k_H, given by

$$k_H = \frac{h}{\rho U_\infty C_p} \equiv N_u \,.\, R_e^{-1} \,.\, P_r^{-1} \quad (A.18)$$

The main advantage of using k_H is that it is closely linked with C_f, the skin friction coefficient, by

$$k_H = \tfrac{1}{2} S C_f \quad (A.19)$$

where S is the " Reynolds Analogy Factor " which is, approximately, dependent only on the physical properties of the fluid through P_r, the Prandtl number. For both laminar and turbulent incompressible flow

$$S \simeq P_r^{-2/3} \quad (A.20)$$

which leads to eqn. (A.17) above since for local skin friction in laminar flow

$$C_f = 0{\cdot}664 R_e^{-1/2} \quad (A.21)$$

The variation of P_r with temperature for air is quite small,[120] viz. $0{\cdot}68 < P_r < 0{\cdot}74$ in the temperature range above 200°K.

For compressible flow the Nusselt number shows a dependence on Mach number which would involve variations in the constant G in (A.16). However, such variations are obviated if all the fluid properties appearing in the incompressible flow formulae are

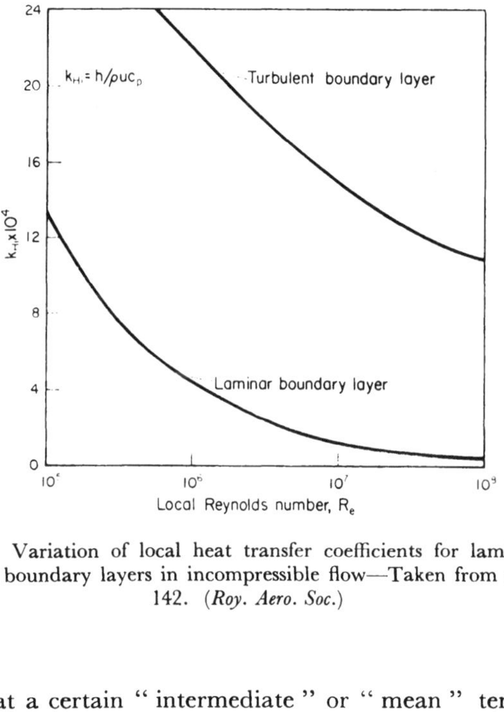

FIG. A.1. Variation of local heat transfer coefficients for laminar and turbulent boundary layers in incompressible flow—Taken from reference 142. (*Roy. Aero. Soc.*)

evaluated at a certain " intermediate " or " mean " temperature. This is discussed more fully later.

The values of the heat transfer coefficients to be used in particular circumstances will not be considered further here. For extensive discussions and analyses appropriate to high speed flow reference should be made to papers by Eckert,[34,35] Monaghan[120] and Seddon.[150] To illustrate the form of the results, Fig. A.1 from reference 142 presents variations with Reynolds number of the incompressible values of Stanton number (k_{Hi}), for both laminar and turbulent flows over a flat plate. The effect of compressibility is only significant in the turbulent case and is shown in Fig. A.2 where the ratio k_H/k_{Hi} is plotted against the temperature ratio T'_S/T'_∞ where T'_S is the surface temperature and T'_∞ is the free stream fluid temperature, both in degrees absolute. In using k_{Hi} the quantities ρ, U_∞ and C_p are evaluated for the aerodynamic conditions just outside the boundary layer.

The representative temperature difference ΔT to be used in low

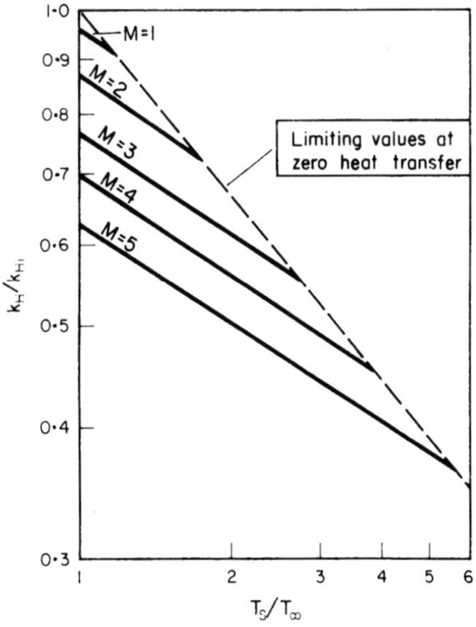

FIG. A.2. Variation of local heat transfer coefficient for turbulent boundary layers in compressible flow—Taken from reference 142. (*Roy. Aero. Soc.*)

speed flow is taken to be $(T_\infty - T_S)$, i.e. that existing between the uniform fluid temperature outside the temperature boundary layer (T_∞) and the surface of the body (T_S). For high speed flow there may be considerable variation of temperature in the boundary layer and instead of T_∞ a suitable " forcing " temperature for the fluid is that temperature which the surface would have when no heat is being transferred to the body. This temperature is known as the " adiabatic wall temperature ", T_{AW}, and it is related to the stagnation temperature.

For a fluid with a velocity U_∞ relative to a body at rest the stagnation temperature is defined as the temperature the fluid would attain if it were brought to rest on the body without addition of heat or external work. The total energy of the fluid (per unit mass) at the outer edge of the boundary layer when equated to the value at the stagnation point on the solid surface gives

$$JC_pT'_\infty + \tfrac{1}{2}U_\infty^2 = JC_pT'_H \tag{A.22}$$

where J is the mechanical equivalent of heat, (1400 ft. lb./C.H.U.) and the specific heat is assumed constant for the temperature range T'_∞ to T'_H. Equation (A.22) may be rewritten as,

$$T'_H = T'_\infty\left[1 + \frac{\gamma - 1}{2}M_\infty^2\right] \tag{A.23}$$

where γ = ratio of specific heats C_p/C_v; M_∞ = Mach number. For air $\gamma = 1{\cdot}4$ and

$$T'_H - T'_\infty \simeq (U_\infty/100)^2 \tag{A.24}$$

where the rise in temperature of the air is in °C when the flow velocity is in miles per hour. It is seen therefore that the maximum temperature of the fluid may be very much greater than the free stream value. This high temperature fluid at the body surface constitutes for aircraft the problem known as *aerodynamic heating*.

If the temperature variations in the fluid are large C_p may not be assumed constant and eqn. (A.22) becomes,

$$Ji_H = Ji_\infty + U_\infty^2/2 \tag{A.25}$$

where i_H is the stagnation enthalpy equal to $\int_0^{T'_H} C_p \mathrm{d}T'$. In this case, (A.12) is rewritten as

$$q = h'\,(i_{AW} - i_S) \tag{A.26}$$

where h' has the units of h/C_p and i_{AW} is defined by

$$i_{AW} = i_\infty + rU_\infty^2/2J \tag{A.27}$$

where i_{AW} is termed the *recovery enthalpy* corresponding to the *recovery temperature*, T'_{AW} (otherwise known as the "zero heat transfer temperature" or "adiabatic wall temperature"), and r is termed the *recovery factor*. The values i_{AW} and T'_{AW} are associated with the layer of fluid adjacent to the body surface if there is no heat transfer at the surface. The recovery factor, r, is defined as

$$r = (i_{AW} - i_{\infty})/(i_H - i_{\infty}) \tag{A.28}$$

and for air it is slightly less than unity. Analyses for both laminar and turbulent flow give, with $P_r = 0{\cdot}72$ for air

$$\left.\begin{aligned} r &\simeq P_r^{1/2}\text{—laminar} \simeq 0{\cdot}85 \\ r &\simeq P_r^{1/3}\text{—turbulent} \simeq 0{\cdot}90 \end{aligned}\right\} \tag{A.29}$$

Values of the zero heat transfer temperature for various flight conditions are given in Table A1.

TABLE A1. VALUES OF ZERO HEAT TRANSFER TEMPERATURE FOR VARIOUS CONDITIONS[120]

Condition		Mach No. 1	2	3	4	5
altitude 0 ft $T_{\infty} = 15°C$						
T_{AW}	laminar	64	206	428	725	1086
°C	turbulent	66	216	456	774	1160
altitude 25,000 ft $T_{\infty} = -34{\cdot}5°C$						
T_{AW}	laminar	6	124	313	565	878
°C	turbulent	7	132	336	607	937
altitude 36,000 ft $T_{\infty} = -56{\cdot}5°C$						
T_{AW}	laminar	−20	88	260	493	780
°C	turbulent	−18	95	282	532	834

To avoid the necessity of changing G in eqn. (A.16) with variations in Mach number it is the usual practice to base the estimates of heat transfer coefficient on the proposals of Eckert[34] or Monaghan.[120] These methods, the "intermediate" and "mean" enthalpy methods, follow from a suggestion made by Young and Janssen[174] that formulae for estimating skin friction coefficients in incompressible flow can also be used for compressible flow if the Reynolds number is appropriately modified. The procedure is to evaluate all the fluid properties at a certain intermediate (Eckert) or weighted average (Monaghan) temperature or enthalpy and it is then found that compressible and incompressible results correlate well without the necessity of changing the constant, G, in (A.16) with variations in Mach number. Thus, Eckert's " intermediate " or " reference " enthalpy is,

$$i_R = i_\infty + 0{\cdot}50\,(i_S - i_\infty) + 0{\cdot}22\,(i_{AW} - i_\infty) \qquad \text{(A.30)}$$

and Monaghan's " mean " enthalpy is,

$$i_M = i_\infty + 0{\cdot}54\,(i_S - i_\infty) + 0{\cdot}16\,(i_{AW} - i_\infty) \qquad \text{(A.31)}$$

In both cases i_S is the enthalpy evaluated at the body surface temperature.

The above procedure has been confirmed for flat plates with either laminar or turbulent boundary layers over a wide range of supersonic Mach numbers, surface and free stream temperatures. It should be noted that since q depends through h', and hence i_S, on i_R (or i_M) an iterative process is involved in any analysis using i_R (or i_M) to find q unless i_S is initially specified, e.g. as for surfaces which are cooled or ablating.

A.3. Heat Radiation

All bodies at temperature emit radiant electro-magnetic energy at a rate which is found to be dependent on the fourth power of the absolute temperature of the body. When this radiant energy reaches another body it may be either transmitted, reflected or absorbed. The corresponding terms transmissivity, reflectivity and absorptivity denote the fractions of the incident energy so that the sum of the terms is unity. For most engineering materials which are opaque the transmissivity is negligible and the limiting case of unit absorptivity and zero reflectivity is the ideal case of a *black body.*

The total radiant energy emitted in unit time from unit area of a black body into a solid angle of 2π is

$$W_B = \sigma_B\,(T')^4 \qquad \text{(A.32)}$$

This result was first proved from thermodynamic principles (Boltzmann) and justified experimentally (Stefan) and is known as the *Stefan–Boltzmann law*. The Stefan–Boltzmann constant has the value

$$\sigma_B = 2{\cdot}78 \times 10^{-12} \text{ C.H.U./ft}^2 \text{ sec } (^\circ\text{K})^4$$

A *non-black* body at temperature T' is said to have a total hemispherical emissivity ϵ if the radiant energy emitted is

$$W = \epsilon\sigma_B(T')^4 \qquad \text{(A.33)}$$

where ϵ may be a function of surface finish and surface temperature.[170]

By considering the thermal equilibrium of several bodies in radiant energy interchange it can be shown that the ratio of the emissive power of a surface to its absorptivity is the same for all bodies. This is known as *Kirchhoff's law*.

It is seen therefore that elements of a structure may radiate energy either into space or to each other. If two surfaces forming part of a structure are separated by a non-absorbing medium the rate of heat transfer between them may be expressed by,

$$q = K_1(T_1')^4 - K_2(T_2')^4 \qquad \text{(A.34)}$$

where the constants depend upon the emitting and absorbing characteristics of the two surfaces, and their relative disposition. These latter effects may be combined in a single quantity known as the *geometrical view factor*, F, which has been evaluated for various configurations (see references 58, 78 and 111).

It is intuitively obvious that from a thermal stress point of view internal radiant energy interchange within a structure is more significant for structural materials having a low thermal conductivity as radiation may then considerably diminish the magnitude of the temperature gradients within the structure.[68]

Another radiation term which may in some cases be significant is that due to incident radiant energy from solar, terrestrial or stellar sources. Of these, solar radiant energy is the most significant and may be written

$$q = a_s G_s \cos\phi$$

This can be considered as a prescribed heat flux at the body surface where a_s is the surface absorptivity, ϕ is the angle between the normal to the surface and the incident rays, and G_s is the energy "constant", which varies with distance from the sun. In the vicinity of the earth G_s increases with altitude from 100 C.H.U./ft^2 hr at sea level to 240 C.H.U./ft^2 hr at 50,000 ft.[90]

A.4. Equilibrium Solutions

It follows from (A.8) that if H is independent of time the steady state temperature distribution satisfies the equation

$$k\,\nabla^2 T + H = 0 \qquad \text{(A.36)}$$

When $H = 0$ the equilibrium solution for the temperature distribution satisfies the Laplace equation

$$\nabla^2 T = 0 \qquad \text{(A.37)}$$

For a thin plate element of thickness d it may, in many practical problems, be assumed that the temperature distribution is two dimensional only, i.e. in the plane of the plate. Thus the transient heating equation would have the form,

$$Cd\frac{\partial T}{\partial t} = \Sigma q \qquad \text{(A.38)}$$

where Σq would be the total heat input due to conduction along the plane, with convection and radiation normal to the plane. If conduction effects are now neglected eqn. (A.38) may be written in more detail as

$$Cd\frac{\partial T_s}{\partial t} = h'(i_{AW} - i_s) + a_s G_s \cos\phi - \epsilon\sigma_B(T_s')^4 \qquad \text{(A.39)}$$

when heat transfer on only one surface of the plate is considered. In other words, when conduction in a solid body can be neglected (as for a thin plate) the boundary conditions of heat flux enter directly into the transient heating equation (cf. with Section A.5.2). The equilibrium solution corresponds to $\partial T_s/\partial t = 0$ when there is a balance between the heat inputs and outputs for the structure. If T_e' is the equilibrium temperature in degrees absolute (A.39) becomes

$$h'(i_{AW} - i_e) + a_s G_s \cos\phi = \epsilon\sigma_B(T_e')^4 \qquad \text{(A.40)}$$

An iterative approach may be required to solve (A.40) because of the possible dependence of ϵ on T_e' and h' on i_e. Note that T_e is usually less than T_{AW} (See Table A2.)

TABLE A2. EFFECT OF EMISSIVITY ON EQUILIBRIUM TEMPERATURES ATTAINED IN SUSTAINED FLIGHT AT $M = 3{\cdot}0$ AND $M = 4{\cdot}0$ AT 70,000 ft

ϵ	T_e °C Equilibrium temperature			
	laminar		turbulent	
	$M = 4{\cdot}0$	$M = 3{\cdot}0$	$M = 4{\cdot}0$	$M = 3{\cdot}0$
0	493	260	532	282
0·15	345	194	475	267
0·5	244	133	405	240
1·00	187	93	353	214

Equilibrium solutions may or may not give rise to severe thermal stress conditions depending on the heat flux boundary conditions assumed at different parts of the body surface. However, it is necessary to evaluate T_e at different parts of the structure since, as the maximum temperature the structure can attain, it can influence the selection of the materials in the structure. Tables A3 and A4 illustrate the dependence on temperature of various physical and mechanical properties, respectively, for several materials.

A.5. Finite Difference Formulation of the Heat Transfer Problem

As was stated earlier there are many problems when, because of non-linear boundary conditions (e.g. forced convection) or temperature dependent properties, the available analytical methods for the solution of the heat conduction equations are no longer applicable. Finite difference procedures may be used however, and a typical heat transfer problem will now be considered using such a procedure. The various boundary conditions which may be applied in heat transfer analyses will be presented in their usual and in their finite difference forms.

Table A.3. Physical properties of various materials

Material	Specific gravity	Thermal conductivity CHU/hr ft² °C/ft				Thermal diffusivity ft²/hr				Melting point (°C approx)
		20	100	250	400°C	20	100	250	400°C	
Aluminium alloy D.T.D.683	2·7	80	86	98	110	2·2	2·4	2·8	3·1	660
Titanium alloy I.C.I.314A	4·5	10	10	10	10	0·3	0·23	0·23	0·20	1800
Stainless steel Jethete M 160	7·8	13	14	15	15	0·3	0·3	0·3	0·3	1400
Epoxy-glasscloth	0·1	1	—	—	—	0·05	—	—	—	600
Magnesium alloy Z5Z	1·8	65	—	—	—	2·6	—	—	—	650

TABLE A4. MECHANICAL PROPERTIES OF VARIOUS MATERIALS

Material	Specific ultimate strength $(ton/in^2)/(lb/in^3)$				Specific stiffness $(lb/in^2)/(lb/in^3) \times 10^{-6}$				Thermal stress modulus $E\alpha$ lb/in^2 °C			
	20	100	250	400°C	20	100	250	400°C	20	100	250	400°C
Al alloy	335	300	90	20	95	90	70	50	220	210	175	135
Hiduminium 100	230	180	130	80	105	95	80	65	—	—	—	—
Titanium alloy	380	350	290	280	95	92	85	80	157	155	140	128
Stainless steel	210	210	200	150	110	105	100	92	270	290	310	300
Epoxy glasscloth	280	250	160	10	65	50	35	0	8	6	4	0
Magnesium alloy	260	190	90	0	95	76	50	0	180	140	90	0

For a more extensive treatment of this subject reference should be made to Dusinberre,[33] Hildebrand,[66] Kosko[98] or Schuh.[149]

A.5.1. The Heat Conduction Equation

The transient temperature distribution for one-dimensional heat conduction along a body of uniform cross-section is described by,

$$C\frac{\partial T}{\partial t} = \frac{\partial}{\partial x}\left(k\frac{\partial T}{\partial x}\right) \tag{A.41}$$

where the material properties, k and C, may be temperature dependent. In finite difference form this equation becomes

$$\left.\begin{aligned} C_{m,n+1/2}\frac{(T_{m,n+1} - T_{m,n})}{\Delta t} &= \frac{1}{\Delta x}\left[k_{m-1/2}\frac{(T_{m-1,n} - T_{m,n})}{\Delta x}\right. \\ &\qquad \left. - k_{m+1/2}\frac{(T_{m,n} - T_{m+1,n})}{\Delta x}\right] \end{aligned}\right\} \tag{A.42}$$

where m and n refer, respectively, to the number of spatial and time intervals. Here, Δx is the spatial interval (assumed constant), Δt is the time interval and, to allow for temperature-dependence, the thermal properties are evaluated at the appropriate temperatures. Thus $C_{m,n+1/2}$ is the heat capacity per unit volume of the material in element m at the mean temperature in the time interval n to $n + 1$ (i.e. $[T_{m,n+1} + T_{m,n}]/2$); $k_{m-1/2}$ and $k_{m+1/2}$ should likewise be evaluated at time n, at the mean temperatures of elements $m - 1$ and m, and $m + 1$ and m, respectively (i.e. $[T_{m-1,n} + T_{m,n}]/2$ and $[T_{m,n} + T_{m+1,n}]/2$).

In eqn. (A.42) the *forward difference* formulation has been used for the heat storage term $C\,\partial T/\partial t$ and, in general, the solution of equations of this type must be iterative because of the term $C_{m,n+1/2}$. For C independent of temperature, or if it may be evaluated with sufficient accuracy at the temperature $T_{m,n}$ (i.e. $C_{m,n+1/2} \simeq C_{m,n}$), eqn. (A.42) may be rearranged to give $T_{m,n+1}$ directly in terms of known quantities. In such a case the solution of the set of equations of type (A.42) is particularly straightforward. Other formulations are of course possible, e.g. the *central difference* formulation yields

$$C\frac{\partial T}{\partial t} \equiv C_{m,n}\frac{(T_{m,n+1} - T_{m,n-1})}{2\Delta t} \tag{A. 43}$$

where $C_{m,n}$ is evaluated at $T_{m,n}$; and the *backward difference* formulation gives,

$$C\frac{\partial T}{\partial t} \equiv C_{m,n-1/2}\frac{(T_{m,n} - T_{m,n-1})}{\Delta t} \tag{A.44}$$

where $C_{m,n-1/2}$ is evaluated at the mean temperature

$$(T_{m,n} + T_{m,n-1})/2.$$

However, the choice of formulation is not arbitrary and each type should be examined to determine whether the solutions obtained are satisfactory. Schuh[149] has shown that four different types of behaviour need to be considered, viz.

(*a*) Divergent solutions must be prevented.

(*b*) Oscillatory solutions must be prevented.

(*c*) The temperatures at any time must be such that heat does not flow in the direction of increasing temperature, i.e. $\mathrm{d}T/\mathrm{d}x$ must be of opposite sign to the heat flow.

(*d*) The balance of heat flow for a section must have the same sign as the time derivative of temperature for the section. i.e. if incoming heat is greater than outgoing heat, then $\mathrm{d}T/\mathrm{d}t > 0$.

Because of these considerations it can be shown that limitations must be placed on the relative sizes of the spatial and time intervals, Δx and Δt. Thus, for a material with constant thermal properties, (A.42) may be rewritten as,

$$T_{m,n+1} = T_{m,n}(1 - 2R) + R(T_{m-1,n} + T_{m+1,n}) \tag{A.45}$$

where $R = \kappa\Delta t/(\Delta x)^2$.

From consideration (*b*) it follows that if all temperatures are initially zero except at one point where $T_{m,n} > 0$, any subsequent negative values of temperature are not physically permissible. Thus if $T_{m-1,n}$ and $T_{m+1,n}$ are both zero the condition that $T_{m,n+1} \geqslant 0$ is that $R \leqslant \frac{1}{2}$. Since R must be positive by definition, it may be seen that

$$0 < R \leqslant \tfrac{1}{2} \tag{A.46}$$

Similar studies by Schuh[149] have shown that considerations (*c*) and (*d*) lead, respectively, to upper limits on R of $\frac{1}{3}$ and $\frac{1}{4}$.

It has also been proved that the error of the finite difference procedure of eqn. (A.45) is a minimum for $R = \frac{1}{6}$.

The limitations on the upper limit of R imply that for a given Δx

the time interval Δt also has an upper limit. For a reasonably fine mesh a large number of small time intervals would therefore be required in a given problem. It should be noted that for the maximum permissible value of $R = \frac{1}{2}$ a particularly simple equation is obtained from (A.45), viz.

$$T_{m,n+1} = (T_{m-1,n} + T_{m+1,n})/2 \qquad \text{(A.47)}$$

Thus the temperature at a point is the mean of the temperatures at adjacent points at the previous time interval. This is the counterpart of Schmidt's averaging graphical procedure.[98]

The stability of finite difference solutions has also been examined by Gaumer[52] who has presented results based on *forward*, *central* and *backward* difference formulations for the heat transfer terms, i.e.

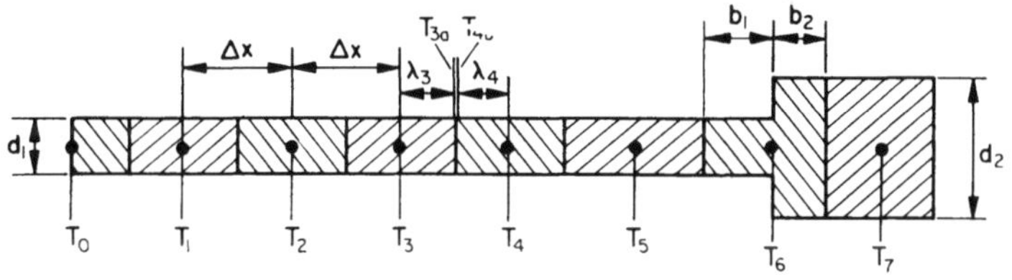

FIG. A.3. One-dimensional mesh for finite difference formulation.

conduction — $\partial T/\partial x$, convection and radiation; but with the heat storage term, $\partial T/\partial t$, formulated as in eqn. (A.42). The results obtained show that the effect of the heat transfer terms on the boundary, i.e. convection and radiation, may be to impose more severe limitations on the upper limit of R.

To select the elements to be used in the finite difference formulation, the usual procedure is to locate temperature points at all boundaries and then locate additional points (as many as possible) within the solid. The length of each heat balance element need not be constant, as indicated in Fig. A.3. which illustrates a typical spatial mesh for a one-dimensional formulation.

A.5.2. Boundary Conditions

I. Prescribed surface temperature: In general this may be written as

$$T_{(s,t)} = F_{(s,t)} \qquad \text{(A.48)}$$

where F is a prescribed function to be satisfied at all points on the surface s and at all times t. In the finite difference formulation this condition is satisfied by taking a control point, say 0, at the surface (see Fig. A.3) and the temperature $T_{(0,n)}$ is entered directly into the calculations for $T_{1,n+1}$ in eqn. (A.45). For a step change in temperature $T_{0,0}$ at the surface, Dusinberre[33] recommends that the mean value of $T_{0,0}$ be used at time zero in the calculation for $T_{1,1}$—with the final value of $T_{0,n}$ thereafter. This improves the accuracy of the solution.

II. Prescribed heat input: This condition is written as

$$k\frac{\partial T_{(s,t)}}{\partial n} = q_{(s,t)} \tag{A.49}$$

where n is the outward normal at the point s and q is the prescribed function. In the finite difference formulation the value of q is entered directly into the calculations for the surface point, e.g. point 0 in Fig A.3. For the end element of length $\Delta x/2$ a heat balance analysis gives,

$$q\Delta t + k\Delta t\,(T_{1,n} - T_{0,n})/\Delta x = \frac{C\Delta x}{2}(T_{0,n+1} - T_{0,n}) \tag{A.50}$$

or

$$T_{0,n+1} = T_{0,n}\,(1 - 2R) + 2RT_{1,n} + 2Rq\Delta x/k \tag{A.51}$$

It follows from (A.49) that for a *perfectly insulated surface*, i.e. one across which there is no heat flux, q,

$$\frac{\partial T_{(s,t)}}{\partial n} = 0 \tag{A.52}$$

At such a surface, e.g. at the mid-point m of a symmetrically heated slab, $T_{m+1,n} = T_{m-1,n}$. Therefore eqn. (A.45) becomes

$$T_{m,n+1} = T_{m,n}\,(1 - 2R) + 2RT_{m+1,n} \tag{A.53}$$

the same result is found from (A.51) with $q = 0$.

III. Convection condition: From (A.12) this may be written as

$$k\frac{\partial T_{(s,t)}}{\partial n} = h_{(s,t)}\,[T_{AW(s,t)} - T_{(s,t)}] \tag{A.54}$$

or even more generally, (following A.26), as

$$k \frac{\partial T_{(s,t)}}{\partial n} = h'_{(s,t)} \; [i_{AW(s,t)} - i_{(s,t)}] \tag{A.55}$$

Denoting the right-hand side of (A.54) by q and substituting into (A.51), with $T_{(s,t)} = T_{0,n}$, yields,

$$T_{0,n+1} = T_{0,n} \, [1 - 2R(1 + B)] + 2RT_{1,n} + 2RBT_{AW} \tag{A.56}$$

where $B = h\Delta x/k$ has the form of a Biot number.

Stability consideration (*b*) of Section A.5.1 produces the following upper limit on R, viz.

$$R \leqslant 1/2(1 + B) \tag{A.57}$$

Thus the convection boundary condition imposes a more severe limitation on the upper limit of R, than does an internal element for which $R \leqslant 1/2$. This result was mentioned earlier and it can be seen that the effect of the term B can be quite important. In order to keep B small and accordingly its effect on R, for a given material and heat transfer condition requires Δx to be small. This has the further effect of making Δt even smaller from the definition of R.

Gaumer[52] showed that this severe limitation is removed if the problem is formulated with the heat transfer terms expressed in *backward differences*. Unfortunately this results in a set of simultaneous equations for the various $T_{m,n+1}$ whereas the *forward difference* approach results in equations each having only one unknown.

IV. Radiation condition : The general result may be written from eqn. (A.34), viz.

$$k \frac{\partial T_{(s,t)}}{\partial n} = q = K_R \, (T'_R)^4 - K_s \, (T'_s)^4 \tag{A.58}$$

This expression for q may be substituted directly into eqn. (A.51) with T'_s replaced by $T'_{0,n}$ and T'_R becoming $T'_{R,n}$—the absolute temperature of a radiative heat source. If other than the *forward difference* formulation is used for the heat transfer terms, the following approximations[52] are made to maintain first-order simultaneous equations in the *central* and *backward* difference methods respectively,

$$\left.\begin{aligned} (T'_{m,n+1/2})^4 &\simeq 2(T'_{m,n})^3 \, T'_{m,n+1} - (T'_{m,n})^4 \\ (T'_{m,n+1})^4 &\simeq 4(T'_{m,n})^3 \, T'_{m,n+1} - 3(T'_{m,n})^4 \end{aligned}\right\} \tag{A.59}$$

V. Solid surfaces in contact: If the interfacial surfaces of two bodies are in perfect thermal contact the necessary boundary conditions are that the temperatures and heat flows in both bodies must be equal at the interface. Thus,

$$T_{a(s,t)} = T_{b(s,t)} \tag{A.60}$$

$$k_a \frac{\partial T_{a(s,t)}}{\partial n} = k_b \frac{\partial T_{b(s,t)}}{\partial n} \tag{A.61}$$

where subscripts a and b refer to the two bodies having a common normal n at s.

The concept of *joint thermal conductance* is introduced when the thermal contact is imperfect, which defines a proportionality between the heat flow and the temperature difference of the two surfaces. Thus, (A.60) is replaced by

$$k_a \frac{\partial T_{a(s,t)}}{\partial n_a} = h_j \left[T_{b(s,t)} - T_{a(s,t)} \right] \tag{A.62}$$

where n_a is the outward normal referred to surface " a " at point s. The joint thermal conductance, h_j, is a function of many parameters and, in general, no specific values can be ascribed to it. Experimentally determined values based on similar structural configurations should be used in any analytical study and where a range of values is possible (in fact this is probable) extreme values should be considered to indicate the effects on the temperature distribution, and hence the thermal stress distribution, in the structure. The results of several experimental investigations are summarized below for the thermal conductance of various aircraft-type joints.[9] The results quoted below are based on tests on 100 specimens incorporating either riveted, screwed or spot-welded joints, i.e.

(1) h_j varied from 80–2150 CHU/ft^2 hr °C.
(2) h_j varied considerably for " identical " specimens.
(3) h_j not affected by interface cleanliness.
(4) h_j increases with interface pressure.
(5) h_j varies considerably with material combination.
(6) h_j varies considerably with type of joint connexion.

N.B. Heat transfer across a joint is given by

$$q = h_j \left(T_a - T_b \right)$$

Heat transfer across a joint must depend upon the degree of flatness of adjacent surfaces, upon the pressure across the joint and the surface roughness—these may be called *manufacturing* effects. The type of joint and the materials used are also important and attributable to the *design*. The fact that h_j depends upon the value of the heat transfer itself and the mean temperature in the joint shows the effect of *environment*.

If eqn. (A.62) is assumed to apply to the one-dimensional system of Fig. A.3 it may be used directly. Thus, if T_{3a} and T_{4b} denote the interface temperatures at a joint between elements 3 and 4, eqns. (A.61) and (A.62) become

$$k_3(T_3 - T_{3a})/\lambda_3 = h_j(T_{3a} - T_{4b}) = k_4(T_{4b} - T_4)/\lambda_4 \tag{A.63}$$

where λ_3 and λ_4 are half the respective element lengths. Since the quantities T_{3a} and T_{4b} would not normally enter into the finite difference formulation in the absence of a joint it may be thought that two additional unknown quantities have been added to the problem. However, expressions for T_{3a} and T_{4b} can be found from (A.63) in terms of the other quantities and the interface temperatures are effectively removed from the problem by defining an effective conductance between temperature points T_3 and T_4 at the centres of the respective elements. Thus,

$$H_j(T_3 - T_4) = h_j(T_{3a} - T_{4b}) \tag{A.64}$$

and

$$1/H_j = 1/h_j + \lambda_3/k_3 + \lambda_4/k_4 \tag{A.65}$$

Therefore the overall *thermal resistance* from one element centre to the next, across a joint, is obtained by summing the individual resistances in series. This analogy with electrical resistance is the basis of electrical analogues[102, 136] which are used to solve many difficult problems of transient heat transfer.

In this respect it should be noted that pneumatic and hydraulic analogues have also received considerable development for application to heat transfer problems.[21, 95, 101, 121] The advantages of such analogues are that only a spatial mesh is required as the time derivative is retained; also, various non-linearities can be introduced with very little extra complication, e.g. variation of specific heat with temperature, latent heat effects, etc.

A.5.3. Some Typical Formulations

To utilize the results previously obtained consider the typical thin sheet structures shown in Figs. A.4 and A.5.

From Fig. A.4. it is seen that a mechanical joint (rivet, bolt, etc.) joins elements 6 to 12 and 7 to 13. If convective heat transfer is assumed on the upper faces of the skin elements only, the transient temperature variations for the different elements are determined from the typical equations below (radiation terms not included).

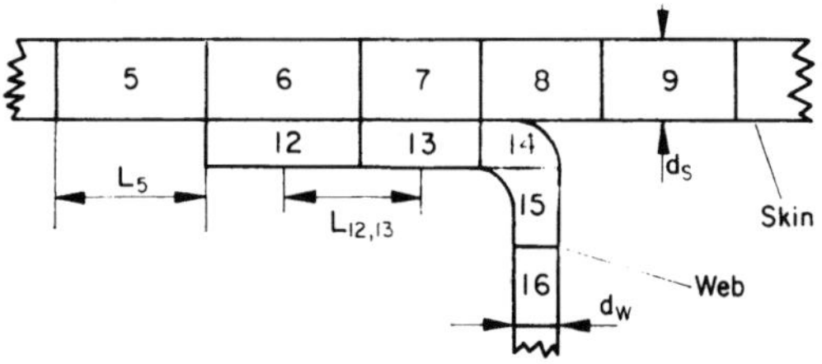

FIG. A.4. Typical thin sheet structure with mechanical joint.

Skin element 5:

$$\frac{L_5 d_s C_5}{\Delta t}(T_{5,n+1} - T_{5,n}) = h_5 L_5 (T_{AW,n} - T_{5,n}) + \frac{k_{4,5} d_s}{L_{4,5}}(T_{4,n} - T_{5,n}) + \frac{k_{5,6} d_s}{L_{5,6}}(T_{6,n} - T_{5,n}) \quad \text{(A.66)}$$

Skin element 7 (with joint):

$$\frac{L_7 d_s C_7}{\Delta t}(T_{7,n+1} - T_{7,n}) = h_7 L_7 (T_{AW,n} - T_{7,n}) + \frac{k_{6,7} d_s}{L_{6,7}}(T_{6,n} - T_{7,n}) + \frac{k_{7,8} d_s}{L_{7,8}}(T_{8,n} - T_{7,n}) + H_{j7} L_7 (T_{13,n} - T_{7,n}) \quad \text{(A.67)}$$

Web element 15:

$$\frac{L_{15} d_w C_{15}}{\Delta t}(T_{15,n+1} - T_{15,n}) = \frac{k_{14,15}}{L_{14,15}}(T_{14,n} - T_{15,n}) d_w + \frac{k_{15,16}}{L_{15,16}}(T_{16,n} - T_{15,n}) d_w \quad \text{(A.68)}$$

Apart from the geometrical parameters (L, d_s, d_w) which are defined in Fig. A.4 the remaining terms have all been defined previously. The quantities C, k, H_j, h may all be complicated functions of temperature and the suffices used have the obvious significance, e.g. as defined after (A.42).

The limiting case of zero joint resistance ($h_j = \infty$) enables eqn. (A.65) to be rewritten as

$$H_j = k_3k_4/(\lambda_3k_4 + \lambda_4k_3) \tag{A.69}$$

In this case, however, the last term in eqn. (A.67) is more easily expressed by $k_{7,13}\ (T_{13,n} - T_{7,n})L_7/L_{7,13}$, when elements 7 and 13 are of similar material (otherwise (A.69) should be used).

The structural configuration shown in Fig. A.5 corresponds to integral construction or welded construction and would have zero joint thermal resistance. The equations to be used are the same as

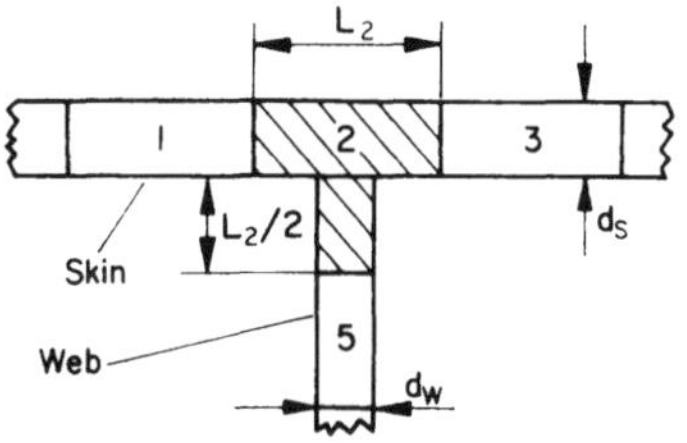

FIG. A.5. Typical thin sheet structure with integral construction.

those above, except for element 2 which is considered to consist of a length of skin L_2 and a depth of web $L_2/2$. The heat balance equation for this element is

$$\left(L_2d_s + \frac{L_2d_w}{2}\right)C_2\left[\frac{T_{2,n+1} - T_{2,n}}{\Delta t}\right] = h_2L_2(T_{AW,n} - T_{2,n})$$

$$+ \frac{k_{1,2}d_s}{L_{1,2}}(T_{1,n} - T_{2,n}) + \frac{k_{2,3}d_s}{L_{2,3}}(T_{3,n} - T_{2,n}) + \frac{k_{2,5}d_w}{L_{2,5}}(T_{5,n} - T_{2,n}) \tag{A.70}$$

A similar equation may be derived for the case of a body having a step change in cross-section, e.g. at temperature point 6 in Fig A.3. Thus,

$$(b_1d_1 + b_2d_2)C_6\left[\frac{T_{6,n+1} - T_{6,n}}{\Delta t}\right] = \frac{k_{5,6}d_1}{L_{5,6}}(T_{5,n} - T_{6,n})$$

$$+ \frac{k_{6,7}d_2}{L_{6,7}}(T_{7,n} - T_{6,n}) \tag{A.71}$$

where $(b_1 + b_2) = L_6$.

References

1. Abir, D. and Nardo, S. V., Thermal buckling of circular cylindrical shells under circumferential temperature gradients, *J. Aero-space Sci.*, **26**, (12) 803, (1959).
2. Anderson, M. S., Combinations of temperature and axial compression required for buckling of a ring-stiffened cylinder, *NASA TN* D 1224 (1962).
3. Anderson, M. S., Thermal buckling of cylinders, *NASA TN* D 1510 (1962).
4. Anderson, M. S. and Card, M. F., Buckling of ring-stiffened cylinders under a pure bending moment and a non-uniform temperature distribution, *NASA TN* D 1513 (1962).
5. Argyris, J. H. and Kelsey, S., *Energy Theorems and Structural Analysis*, Butterworths, London, 1960. Also in *Aircraft Engng.*, (1954–55).
6. Argyris, J. H and Kelsey, S., *Modern Fuselage Analysis and the Elastic Aircraft*, Butterworths, London, 1963. Also in *Aircraft Engng.*, (1959 and 1961).
7. Ayers, K. B., Thermal stresses in I-section beams, *Aircraft Engng.*, **34** (11) 320 (1962).
8. Barker, L., The calculation of temperature stresses in tubes, *Engineering*, **124**, 443 (1927).
9. Barzeley, M. E., Range of interface thermal conductance for aircraft joints, *NASA TN* D 426 (1960).
10. Batdorf, S. B., A simplified method of elastic stability analysis for thin cylindrical shells, *NACA* TR. 874 (1947).
11. Bijlaard, P. P., Differential equations for cylindrical shells with arbitrary temperature distributions, *J. Aero-space Sci.*, **25**, (9) 594 (1958).
12. Bijlaard, P. P., Thermal stresses and deflections in rectangular sandwich plates, *J. Aero-space Sci.*, **26**, (4) 210 (1959).
13. Bisplinghoff, R. L., Further remarks on the torsional rigidity of thermally stressed wings, *J. Aero-space Sci.*, **25**, (10) 657 (1958).
14. Boley, B. A., The determination of temperatures, stresses and deflections in two-dimensional thermoelastic problems, *J. Aero. Sci.*, **23**, (1) 67 (1956).
15. Boley, B. A. and Weiner, J., *Theory of Thermal Stresses*, John Wiley, New York, 1960.
16. Broglio, L. and Santini, P., Thermal stresses, Chap. 10, *High Temperature Effects in Aircraft Structures*, Agardograph No. 28, Pergamon Press, 1958.
17. Buckens, F., Dynamic behaviour of plates under thermal stress, Univ. of Louvain (Belgium) Rep. AF.61(052)–499 TRI March (1962).
18. Carslaw, H. S. and Jaeger, J. C., *Conduction of Heat in Solids*, Clarendon Press, Oxford, 1947.
19. Cotterell, B. and Parkes, E. W., Thermal buckling of circular plates, *Brit. Aero. Res. Council*, R and M No. 3245 (1960).
20. Cox, M. and Glenny, E., Thermal fatigue investigations, *Engineer*, **210**, 346 (1960).
21. Coyle, M. B., An air-flow analogy for the solution of transient heat conduction problems, *Brit. J. Appl. Phys.*, **2** (1) 12 (1951).

22. DAS, Y. C. and NAVARATNA, D. R., Thermal bending of rectangular plates, *J. Aero-space Sci.*, **29** (11) 1397 (1962).
23. DEN HARTOG, J. P., Temperature stresses in flat rectangular plates and in thin cylindrical tubes, *J. Franklin Inst.*, **222**, 149 (1936).
24. DENKE, P. H., A matrix method of structural analysis, *Proc. 2nd U.S. Nat. Cong. Appl. Mech.*, (Amer. Soc. Mech. Engrs.) 445, (1954).
25. DE SILVA, C. N., Thermal stresses in the bending of ogival shells, *J. Aero-space Sci.*, **29** (2), 207 (1962).
26. DE VEUBEKE, B. F., Creep buckling, Chap. 13 of *High Temperature Effects in Aircraft Structures*, Agardograph No. 28, Pergamon Press, 1958.
27. DOBBINS, R., The minimum weight of a structure protected against short duration aerodynamic heating by means of thermal insulation, *Inst. Aero. Sci.* Preprint No. 773 (1958).
28. DONNELL, L. H., Stability of thin walled tubes under torsion, *NACA* TR. 479, (1933).
29. DUHAMEL, J. M., Seconde memoire sur les phénomènes thermo-mecaniques, *J. Éc. polyt. Paris*, **15**, 1 (1837).
30. DUHAMEL, J. M., Memoire sur le calcul des actions moléculaires developpées par les changements de température dans les corps solides, *Mém. prés. Acad. Sci. Paris*, **5**, 440 (1838).
31. DUHAMEL, J. M., Memoire sur le mouvement des differents points d'une barre cylindrique dont la température varie, *J. Éc. polyt. Paris*, **21**, 1 (1856).
32. DUKES, W. H., Protection of aircraft structures against high temperatures, *Proc. 2nd Internat. Cong. Aero. Sci.* (Zurich), Pergamon Press, (1960).
33. DUSINBERRE, G. M., *Heat Transfer Calculations By Finite Differences*, International Textbook Co. (Pennsylvania) 1961.
34. ECKERT, E. R. G., Survey of heat transfer at high speeds, *WADC TR* 54–70 (1954).
35. ECKERT, E. R. G., Engineering relations for friction and heat transfer to surfaces in high velocity flow, *J. Aero. Sci.*, **22** (8) 585 (1955).
36. ECKERT, E. R. G. and DRAKE, R. M., *Heat and Mass Transfer*, McGraw-Hill, New York, 1959.
37. FLUGGE, W. F., *Stresses in Shells*, Springer-Verlag, Berlin, 1960.
38. FORRAY, M. J. and ZAID, M., Thermal stresses in a circular bulkhead subjected to a radial temperature variation, *J. Aero. Sci.*, **25** (1) 63 (1958).
39. FORRAY, M. J., Thermal stresses in plates, *J. Aero-space Sci.*, **25** (11) 716 (1958).
40. FORRAY, M. J., Thermal stresses in rings, *J. Aero-space Sci.*, **26** (5) 310 (1959).
41. FORRAY, M. J., Formulas for the determination of thermal stresses in rings, *J. Aero-space Sci.*, **27** (3) 238; (6) 478 (1960).
42. FORRAY, M. J. and NEWMAN, M., Axisymmetric bending stresses in solid circular plates with thermal gradients, *J. Aero-space Sci.*, **27** (9) 717 (1960).
43. FORRAY, M. J. and NEWMAN, M., Bending of circular plates due to asymmetric temperature distributions, *J. Aero-space Sci.*, **28** (10) 773 (1961).
44. FORRAY, M. J. and NEWMAN, M., On the post-buckling behaviour of rectangular plates, *J. Aero-space Sci.*, **29** (6) 754 (1962).
45. FORRAY, M. J. and NEWMAN, M., The post-buckling analysis of heated rectangular plates, *J. Aero-space Sci.*, **29** (10) 1262 (1962).
46. GALLAGHER, R. H., PADLOG, J. and BIJLAARD, P., Stress analysis of heated complex shapes, *J. Amer. Rocket Soc.*, **32** (5) 700 (1962).
47. GATEWOOD, B. E., Determination of inelastic stresses at elevated temperatures by strain analysis, *Amer. Air Force WADC TR.*, 56–16 (1956).

48. GATEWOOD, B. E., *Thermal Stresses*, McGraw-Hill, New York, 1957.
49. GATEWOOD, B. E., Inelastic combined thermal and applied stresses in skin-stringer aircraft structures, *J. Aero-Sci.*, **25** (3) 212 (1958).
50. GATEWOOD, B. E., The problem of strain accumulation under thermal cycling, *J. Aero-space Sci.*, **27** (6) 461 (1960).
51. GATEWOOD, B. E., Inelastic redundant analysis and test data comparison for a heated ring frame, *J. Aero-space Sci.*, **29** (3) 364 (1962).
52. GAUMER, G. R., Stability of three finite difference methods of solving for transient temperatures, *J. Amer. Rocket Soc.*, **32** (10) 1595 (1962).
53. GLASSTONE, S., *Principles of Nuclear Reactor Engineering*, D. Van Nostrand, 1955.
54. GOODIER, J. N., Thermal stress in long cylindrical shells due to temperature variations around the circumference and through the wall, *Canad. J. Res.*, **15A** (4) 49 (1937).
55. GOODIER, J. N., Integration of thermoelastic equations, *Phil. Mag.*, **23**, 1017 (1937).
56. GOODIER, J. N., Thermal stress and deformation, *J. Appl. Mech.*, **24** (3) 467 (1957).
57. GOSSARD, M. L., SEIDE, P. and ROBERTS, W., Thermal buckling of plates, *NACA TN* 2771 (1952).
58. HAMILTON, D. C. and MORGAN, W. R., Radiant interchange configuration factors, *NACA TN* 2836 (1952).
59. HIESLER, M. P., Transient thermal stresses in slabs and circular pressure vessels, *J. Appl. Mech.*, **20**, 261 (1953).
60. HELDENFELS, R. R., The effect of non-uniform temperature distributions on the stresses and distortions of stiffened shell structures, *NACA TN* 2240 (1950).
61. HELDENFELS, R. R., A numerical method for the analysis of stiffened shell structures under non-uniform temperature distributions, *NACA TR* 1043 (1951).
62. HELDENFELS, R. R. and ROBERTS, W. M., Experimental and theoretical determination of thermal stresses in flat plates, *NACA TN* 2769 (1952).
63. HEMP, W. S., Fundamental principles and theorems of thermoelasticity, *Aeronaut. Quart.*, **V11**, 184 (1956).
64. HEMP, W. S., Thermoelastic formulae for the analysis of beams, *Aircraft Engng.*, **28** (11) 374 (1956).
65. HEMP, W. S., Deformation of heated shells, *SUDAER Rep.* 103 *AFOSR TN*-61–770 (1961).
66. HILDEBRAND, F. B., *Introduction to Numerical Analysis*, McGraw-Hill, 1956.
67. HOFF, N. J., Structural problems of future aircraft, *Proc. 3rd Anglo–Amer. Aero. Conf.* 77, *Roy. Aero. Soc.* (1951).
68. HOFF, N. J., Comparison of radiant and conductive heat transfer in a supersonic wing, *J. Aero. Sci.*, **23** (6) 694 (1956).
69. HOFF, N. J., Thermal buckling of supersonic wing panels, *J. Aero. Sci.*, **23** (11) 1019 (1956).
70. HOFF, N. J., Buckling at high temperature, *J. Roy. Aero. Soc.* **61** (11) 756 (1957).
71. HOFF, N. J. Stress distribution in the presence of creep. Chap. 12 of *High Temperature Effects in Aircraft Structures* Agardograph No. 28, Pergamon Press, 1958.
72. HORVAY, G., Transient thermal stresses in circular disks and cylinders, *Trans. Amer. Soc. Mech. Engrs.*, **76** (1) 127 (1954).
73. HOUGHTON, D. S. and CHAN, A. S., Discontinuity stresses at the junction of a pressurised spherical shell and a cylinder, *College of Aeronautics*, Note No. 80 (1958).

74. Houghton, D. S., Discontinuity problems in shell structures, Chap. 10, *Nuclear Reactor Containment Buildings and Pressure Vessels*, Butterworths, 1960.
75. Huth, J. H., Thermal stresses in conical shells, *J. Aero. Sci.*, **20** (9) 613 (1953).
76. Isakson, G., A simple model study of transient temperature and thermal stress distribution due to aerodynamic heating. *J. Aero. Sci.*, **24** (8) 611 (1957).
77. Jaeger, J. C., On thermal stresses in circular cylinders, *Phil. Mag.*, **36**, 418 (1945).
78. Jakob, M. and Hawkins, G. A., *Elements of Heat Transfer and Insulation*, 2nd Ed., John Wiley, New York, 1950.
79. Jakob, M., *Heat Transfer*, John Wiley, New York, Vol. I, 1949, Vol. II, 1957.
80. Johns, D. J., Optimum design of a multicell box subjected to a given bending moment and temperature distribution, *College of Aeronautics*, Note No. 82 (1958).
81. Johns, D. J., Thermal stress distributions in cylindrical shells stiffened by bulkheads or frames for arbitrary temperature distributions, *College of Aeronautics*, Note No. 83 (1958).
82. Johns, D. J., Approximate formulas for thermal stress analysis, *J. Aero-space Sci.*, **25** (8) 524 (1958).
83. Johns, D. J., Comments on " thermal buckling of clamped cylindrical shells ", *J. Aero-space Sci.*, **26** (1) 59 (1959).
84. Johns, D. J., Thermal stress in short uniform cylindrical shells, *Engng. Materials and Design*, **2** (3) 167 (1959).
85. Johns, D. J., Discontinuity stresses in stiffened cylindrical shells. *Aircraft Engng.*, **31** (5) 131 (1959).
86. Johns, D. J., Structural problems at elevated temperatures, *Engng. Materials and Design*, **2** (5) 270 (1959).
87. Johns, D. J., Thermal stresses in a long circular shell with axial temperature variation, *J. Aero-space Sci.*, **27** (5) 393 (1960).
88. Johns, D. J., Houghton, D. S. and Webber, J. P. H., Buckling due to thermal stress of cylindrical shells subjected to axial temperature distributions, *College of Aeronautics* Report No. 147 (1961).
89. Johns, D. J., Local circumferential buckling of thin circular cylindrical shells, *NASA TN* D 1510 (1962).
90. Johnson, H. A., A design manual for determining the thermal characteristics of high speed aircraft, *Amer. Air Force TR*. 5632 (1947).
91. Kent, C. H., Thermal stresses in thin walled cylinders, *Trans. Amer. Soc. Mech. Engrs.*, **53**, 167 (1931).
92. Kitchenside, A. W., The effects of kinetic heating on aircraft structures, *J. Roy. Aero. Soc.*, **62** (2) 105 (1958).
93. Klein, B., A simple method of structural analysis, Parts 1–5, *J. Aero-space Sci.*, **24**, 39, 813 (1957) ; **25**, 385 (1958) ; **26**, 351 (1959) ; **27**, 859 (1960).
94. Klosner, J. M. and Forray, M. J., Buckling of simply supported plates under arbitrary symmetrical temperature distributions, *J. Aero. Sci.*, **25** (3) 181 (1958).
95. Knuth, E. L. and Kumm, E. L., Application of hydraulic analog method to one-dimensional transient heat flow, *J. Amer. Rocket Soc.* (*Jet Propulsion*), **26**, (8) 649 (1956).
96. Kochanski, S. L. and Argyris, J. H., Some effects of kinetic heating on the stiffness of thin wings, Parts 1–2 *Aircraft Engng.*, **29** (10) 310 (1957) ; **30** (2) 32, (3) 82, (4) 114 (1958).

97. Kornecki, A., On the thermal buckling and free vibrations of cylindrical panels heated from the inside, *Bull. Res. Council of Israel*, **11C** (1) 123 (1962).
98. Kosko, E., The numerical determination of transient temperatures in wings, *Canad. Aero. J.*, **3** (3) 87 (1957).
99. Legg, K. L. C. and Stevens, G., Temperature distributions in aircraft structures and the influence of mechanical and physical material properties, *J. Roy. Aero. Soc.*, **62** (3) 174 (1958).
100. Lempriere, B. M., Thermal stresses in a box structure, *College of Aeronautics*, Note No. 84 (1958).
101. Leopold, C. S., Hydraulic analog for the solution of problems of thermal storage, radiation, convection and conduction, *Trans. Amer. Soc. Heat. Vent. Engrs.*, **54**, 389 (1948).
102. Leppert, G., Unsteady state heat conduction, *J. Amer. Soc. Nav. Engrs.*, **64** (3) 611 (1952).
103. Love, A. E., *A Treatise on the Mathematical Theory of Elasticity*, Cambridge Univ. Press, 4th Ed. 1927.
104. Mansfield, E. H., The influence of aerodynamic heating on the flexural rigidity of a thin wing, *Brit. Aero. Res. Council*, R. and M. No. 3115 (1957).
105. Mansfield, E. H., Combined flexure and torsion of a class of thin heated wings, *Brit. Aero. Res. Council*, R. and M. No. 3195 (1958).
106. Mansfield, E. H., Leading edge buckling due to aerodynamic heating, *Brit. Aero. Res. Council*, R. and M. No. 3197 (1959).
107. Mansfield, E. H., Loading and heating of a simple structure with linear work hardening, *Roy. Aircraft Estab. Rep. Struct.* 276 (1962).
108. Mansfield, E. H., Loss of stiffness in a heated wing: an inequality, *J. Roy. Aero. Soc.*, **67** (3) 191 (1963).
109. Marguerre, K., Zur theorie der gekruemmten platte grossen farmaenderung, *Proc. Vth Int. Congr. Appl. Mech.*, 93 (1938).
110. Maulbetsch, J. K., Thermal stresses in plates, *J. Appl. Mech.*, **2**, A141 (1935).
111. McAdams, W. H., *Heat Transmission*, 3rd Ed. McGraw-Hill, 1951.
112. McKenzie, K. I., The buckling of a circular plate with a concentric circular hot spot, *J. Roy. Aero. Soc.*, **64** (2) 105 (1960).
113. McKenzie, K. I., The leading edge buckling of a thin built-up wing due to aerodynamic heating, *Brit. Aero. Res. Council*, R. and M. No. 3295 (1962).
114. McWithey, R., Minimum weight analysis of symmetrical multiweb beam structures subjected to thermal stress, *NASA TN* D 104 (1959).
115. Mendelson, A. and Hirschberg, M., Analysis of elastic thermal stress in thin plates with spanwise and chordwise variations of temperature and thickness, *NACA TN* 3778 (1956).
116. Miller, D. R., Thermal stress ratchet mechanism in pressure vessels, *Trans. Amer. Soc. Mech. Engrs.*, (Series D) **81**, 190 (1959).
117. Mindlin, R. D., Note on the Galerkin and Papkovitch stress functions, *Bull. Amer. Math. Soc.*, **42** (5) 373 (1936).
118. Miura, K., Thermal buckling of rectangular plates, *Aero. Res. Inst. Univ. Tokyo*, Rep. 353 (1960).
119. Miura, K., Thermal buckling of rectangular plates, *J. Aero-space Sci.*, **28** (4) 341 (1961).
120. Monaghan, R. J., Formulae and approximations for aerodynamic heating rates in high speed flight, *Brit. Aero. Res. Council* C.P. 360 (1955).
121. Moore, A. D., The hydrocal, *Industr. Engng. Chem.*, **28** (6) 704 (1936).
122. Myklestad, N. O., Two problems of thermal stress in the infinite solid, *J. Appl. Mech.*, **9** (3) A-131, A-136 (1942).

123. NEAL, B. G. and SYMONDS, P. S., A method for calculating the failure load for a framed structure subjected to fluctuating loads, *J. Instn. Civ. Engrs.*, **35** (3) 186 (1950).
124. NEAL, B. G., Plastic collapse and shakedown theorems for structures of strain-hardening material, *J. Aero. Sci.*, **17** (5) 297 (1950).
125. NEUMANN, F. E., Die gesetze der doppelbrechung des lichts in comprimirten oder ungleichformig erwarmten unkrystallinischen korpern, *Abh. preuss. Akad. Wiss*, Berlin, Zweiter Teil: 1 (1841).
126. NEWMAN, M. and FORRAY, M. J., Bending stresses due to temperature in hollow circular plates. Parts 1–3, *J. Aero-space Sci.*, **27**, (10) 792; (11) 870; (12) 951 (1960).
127. NEWMAN, M. and FORRAY, M. J., Thermal stresses and deflections in thin plates with temperature dependent elastic moduli, *J. Aero-space Sci.*, **29** (3) 372 (1962).
128. NEWMAN, M. and FORRAY, M. J., Axisymmetric large deflections of circular plates subjected to thermal and mechanical loads. *J. Aero-space* Sci., **29** (9) 1060 (1962).
129. NOVOZHILOV, Z., *Theory of Thin Shells*, Noordhoff, Netherlands, (1959).
130. PARKES, E. W., Transient thermal stresses in wings, *Aircraft Engng.*, **25** (12) 373 (1953).
131. PARKES, E. W., Wings under repeated thermal stress, *Aircraft Engng.*, **26** (12) 402 (1954).
132. PARKES, E. W., Incremental collapse due to thermal stress, *Aircraft Engng.*, **28** (11) 395 (1956).
133. PARKES, E. W., Stresses in a plate due to a local hot spot, *Aircraft Engng.*, **29** (3) 67 (1957).
134. PARKES, E. W., A design philosophy for repeated thermal loading, *Agardograph* No. 213 (1958). Presented at AGARD Structures and Materials Panel Meeting, Copenhagen, Denmark, October 1958.
135. PARKES, E. W., Effects of repeated thermal loading—the influence of the variation of strength with temperature on structural behaviour. *Aircraft Engng.*, **32** (8) 222 (1960).
136. PASCHKIS, V. and BAKER, H. D., A method for determining unsteady state heat transfer by means of an electrical analogy. *Trans. Amer. Soc. Mech. Engrs.*, **64** (2) 105 (1942).
137. POHLE, F. V. and OLIVER, H., Temperature distribution and thermal stresses in a model of a supersonic wing, *J. Aero. Sci.*, **21** (1) 8 (1954).
138. POISSON, S. D., Mémoire sur l'équilibre et le mouvement des corps élastiques, *Mém. Acad. Sci. Paris*, **8**, 357 (1829).
139. PRZEMIENIECKI, J. S., Transient temperatures and stresses in plates attained in high speed flight, *J. Aero. Sci.*, **22** (5) 345 (1955).
140. PRZEMIENIECKI, J. S., Transient temperature distributions and thermal stresses in fuselage shells with bulkheads or frames, *J. Roy. Aero. Soc.*, **60** (12) 799 (1956).
141. PRZEMIENIECKI, J. S., Thermal stresses in rectangular plates. *Aeronaut. Quart.*, **X**, 65 (1959).
142. PRZEMIENIECKI, J. S., The design of transparencies, *J. Roy.Aero. Soc.*, **63** (11) 620 (1959).
143. RAMA RAO, K. and JOHNS, D. J., Some thermal stress analyses for rectangular plates. *J. Aero. Soc. of India*, **13** (4) 99 (1961).
144. ROBERTS, E., Elastic design charts for thin plates with spanwise and chordwise variations in temperature, *NASA TN* D 1182 (1962).

145. ROBERTS, J. K., *Heat and Thermodynamics*, 4th Ed. Blackie, 1951.
146. ROBINSON, K., Elastic energy of an ellipsoidal inclusion in an infinite solid, *J. Appl. Phys.*, **22** (8) 1045 (1951).
147. SCHUH, H., On the calculation of temperature distribution and thermal stresses in parts of aircraft structures at supersonic speeds, *J. Aero. Sci.*, **21** (8) 575 (1954).
148. SCHUH, H., Transient temperature distributions and thermal stresses in a skin-shear web configuration at high speed flight, *J. Aero. Sci.*, **22** (12) 829 (1955).
149. SCHUH, H., *Heat Transfer in Structures*, Pergamon Press (To be published).
150. SEDDON, J., Nomogram analysis of heat transfer in supersonic flow, *Aircraft Engng.*, **33** (5) 124 (1961).
151. SINGER, J., Buckling of thin circular conical shells, subjected to axisymmetrical temperature distributions and external pressure, *Israel. Inst. Technology, Haifa*, TN. 3 (1961).
152. SINGER, J., On the equivalence of the Galerkin and Rayleigh–Ritz methods, *J. Roy. Aero. Soc.*, **66** (9) 592 (1962).
153. SPRAGUE, G. H. and HUANG, P. C., Behaviour of aircraft structures under thermal stress, *Trans. Soc. Auto. Engrs.*, **66**, 457 (1958).
154. STERN, M., Analysis of thermal stresses in conical shells, *J. Aero. Sci.*, **22** (7) 506 (1955).
155. STERNBERG, E. and MCDOWELL, E. L., On the steady state thermoelastic problem for the half space., *Quart. Appl. Math.*, **14**, 381 (1957).
156. SUNAKAWA, M. and UEMURA, M., Deformation and thermal stress in a rectangular plate subjected to aerodynamic heating, *Aero. Res. Inst. Univ. Tokyo*, Rep. 359 (1960).
157. SUNAKAWA, M., Deformation and buckling of cylindrical shells subjected to heating, *Aero. Res. Inst. Univ. Tokyo*, Rep. 370 (1962).
158. SWANSON, C. G., VAN DER MAAS, C. J. and HUANG, P. C., Allowable stresses for elevated temperature structures, *Aero. Systems Div.*, ASD-TR-61-8 (1962).
159. SWITZKY, H., FORRAY, M. J. and NEWMAN, M., Thermostructural analysis manual, *WADD* TR 60-517, Part 1, (1962).
160. THOMPSON, A. S., Thermal stress in power-producing elements, *J. Aero. Sci.*, **19** (7) 476 (1952).
161. THOMPSON. A. S. and RODGERS, O. E., *Thermal Power from Nuclear Reactors*, John Wiley, New York, 1956.
162. TIMOSHENKO, S., *Theory of Elastic Stability*, McGraw-Hill, 1936.
163. TIMOSHENKO, S. and GOODIER, J. N., *Theory of Elasticity*, 2nd Ed. McGraw-Hill, 1951.
164. TIMOSHENKO, S. and WOINOWSKY-KRIEGER, S., *Theory of Plates and Shells*, 2nd Ed. McGraw-Hill, 1959.
165. TROTMAN, C. K. and GRIFFIN, K. H., Incremental collapse of a box beam under thermal and mechanical loading, *2nd Int. Conf. on Stress Analysis, Paris* (1962).
166. TSAO, C. H., Thermal stresses in long cylindrical shells, *Trans. Amer. Soc. Mech. Engrs.*, (Series E) **26** (1) 147 (1959).
167. TURNER, M. J., CLOUGH, R. W., MARTIN, H. C. and TOPP, L. J., Stiffness and deflection analysis of complex structures, *J. Aero. Sci.*, **23** (9) 805 (1956).
168. TURNER, M. J., DILL, E. H., MARTIN, H. C. and MELOSH, R. J., Large deflections of structures subjected to heating and external loads, *J. Aero-space Sci.*, **27** (2) 97 (1960).

169. Van der Neut, A., Buckling caused by thermal stress, Chap. 11, *High Temperature Effects in Aircraft Structures*, Agardograph No. 28, Pergamon Press, 1958.
170. Wade, W. R., Measurements of emissivity of certain stably oxidised metals and some refractory and oxide coatings, *NASA Memo*, 1–20–59L (1959).
171. Webber, J. P. H. and Houghton, D. S., Thermal buckling of a free circular plate, *College of Aeronautics*, Note No. 105 (1960).
172. Weiner, J. H. and Mechanic, H., Thermal stresses in free plates under heat pulse inputs, *WADC Tech. Rep.*, 54–428 (1954).
173. Yao, J. C., Thermoelastic differential equations for shells of arbitrary shape, *Amer. Inst. Aero. Astro. J.*, **1** (2) 479 (1963).
174. Young, G. B. W. and Janssen, E., The compressible boundary layer, *J. Aero. Sci.*, **19** (4) 229 (1952).
175. Zuk. W., Thermal buckling of clamped cylindrical shells, *J. Aero. Sci.*, **24** (5) 359 (1957).

BIBLIOGRAPHIES

1. Boley, B. A., Weiner, J. H. and Tolins, I. S., Thermal stress analysis for aircraft structures—Part II, Bibliography. *WADC TR* 56–102. Part II 1955.
This traces the topic of thermal stress from its origins in 1805 up to November 1955 and contains about 250 references.
2. Gallagher, R. H., Krivetsky, A. and Huff, R., Thermal stress determination techniques in supersonic transport aircraft structures—A bibliography of thermal stress analysis references. Bell Aerosystems Co. Report No. 2114-950001 A.S.D. Contract No. AF 33(657)–8936.
This covers the period from 1955–1963 and contains about 600 references.

Index

Printed in Dunstable, United Kingdom

91338445R00130